MW00845676

# Organic Tobacco Growing in America

*The People of Santa Fe Natural Tobacco Company
share a values driven vision:
That our uncompromising commitment to
our natural tobacco products,
the earth from which they come,
the communities on which we depend,
and the people who bring our spirit to life,
is essential to our success.*

# Organic Tobacco Growing
## in America
### and Other Earth-Friendly Farming

by
## Mike Little and Fielding Daniel
## Mark Smith and Jim Haskins

SUNSTONE
PRESS

SANTA FE

© 2008 by Mike Little, Fielding Daniel, Mark Smith and Jim Haskins.
All Rights Reserved.

No part of this book may be reproduced in any form or by any electronic or
mechanical means including information storage and retrieval systems without
permission in writing from the publisher, except by a reviewer
who may quote brief passages in a review.

Sunstone books may be purchased for educational, business or sales
promotional use. For information please write:
Special Markets Department, Sunstone Press,
P.O. Box 2321, Santa Fe, New Mexico 87504-2321.

Book design ✴ Vicki Ahl
Body typeface ✴ Caslon Pro
Printed on acid free paper

---

Library of Congress Cataloging-in-Publication Data

Organic tobacco growing in America and other earth-friendly farming / by
Mike Little ... [et al.].
    p. cm.
  ISBN 978-0-86534-707-6 (softcover : alk. paper)
  1. Tobacco--United States. 2. Organic farming--United States.
I. Little, Mike, 1959-
  SB273.O74 2009
  633.7'1840973--dc22

                        2008046033

---

Published in

**WWW.SUNSTONEPRESS.COM**
SUNSTONE PRESS / POST OFFICE BOX 2321 / SANTA FE, NM 87504-2321 /USA
(505) 988-4418 / ORDERS ONLY (800) 243-5644 / FAX (505) 988-1025

*T*his book is dedicated to organic growers Brownie Van Dorp, William Bing, Allen Ball and others who passed away in the years since organic tobacco in America came to be.

# Contents

# Introduction

Our company began earth-friendly farming practices in the 1980s. Then, as now, our pioneering effort was a natural outgrowth of who we are as a company and a persistent willingness to do things the tobacco industry said could not be done. But doing things differently is who we've always been. All natural tobacco. No additives or flavorings in our tobacco—use of only the best part of the tobacco leaf. No stems, no puffed tobacco, no reconstituted sheet and scrap tobacco. Just pure 100 percent tobacco leaf. It costs us more to produce. But it is well worth it. And has been since the company began in 1982.

As our consumers know from our advertising, no additives in our tobacco does not mean a safer cigarette. Similarly, organically grown tobacco and tobacco manufactured under the organic regime do not produce a safer cigarette. The best thing smokers who are concerned about the health effects of smoking can do is to quit.

In 1989, we took the novel approach of introducing to our growers a program to dramatically reduce the use of pesticides, and we presented them with an unheard of proposition: How about growing pure and natural organic tobacco?

Today, demand for organic tobacco leaf is doubling each year, and the large tobacco companies are looking at, and in some cases adopting, aspects of our reduced pesticide, purity residue clean approach. As of this writing, we have farmers growing certified organic tobacco in North Carolina, Virginia, Tennessee, Kentucky, Ohio, Canada and Brazil. And we are looking to bring many more farmers into the organic and earth-friendly fold.

Why do so? What are the benefits?

Our growers in these pages answer that it is good for the farmer—both financially and environmentally, by reducing and even eliminating the use of many chemicals and the risk of mishandling—and that it is good for the environment, both the natural environment and man-made community.

In the following pages and chapters, we will explain how this is so, often relying on the words of the growers themselves. As with any pioneering effort, much of what we did to bring forward a better way of farming was through trial and error. And we have no illusions about providing the last word on the subject. In many ways, we are just scratching the surface. What we do provide in this book is practical information about what our growers and we have learned over these last 20 years and why we think organic and other earth-friendly growing practices should become mainstream. Our objective is to share with you how it can be done.

Along the way, we have had the pleasure of working with many fine people who share a dedication to producing the finest and most natural tobacco possible. It is a result of that partnership that led us to write this book.

Mike Little
Senior Vice President—Operations

Fielding Daniel
Director of Leaf

# Organic Tobacco Growing: Who and Why?

Just about anyone can grow organic tobacco. But it helps to have a farming background. Many, if not most, of our organic growers previously grew tobacco. Many took their first step away from conventional growing by joining our purity residue clean (PRC) tobacco program. (More about PRC tobacco later.) But a number of new-to-tobacco growers may come from growers of other certified organic products. Accordingly, this book is written with the understanding that the primary reader has at least a basic understanding of farming and, more than likely, tobacco growing.

So with the "who" may likely consider growing organic tobacco established, let's turn to another large question:

Why grow organically?

For many farmers, the answer is that growing organic tobacco is *more profitable*. Simply put, we, who process, manufacture and turn the leaf into finished product, pay growers a lot more for their organic tobacco. As for growing costs, organic does require more labor and following a set of guidelines, but it also requires less conventional "inputs," which saves money.

As you will hear from the growers in their own words, there are other compelling reasons to go organic. Profitability may be high on the list, but so is the shift to sustainable, earth-friendly farming represented by organic. The kinder and gentler method of farming means growing tobacco free of prohibited materials, making the farm more environmentally sound for those who work it.

No one should underestimate the importance of sustainability and proper stewardship of the land.

According to Thomas B. Harding, Jr., founding member and past president of the Organic Trade Association and current board member of other world organic groups, organic farming is critically important to the future of our global community. "Energy may be today's most pressing issue, but water and its growing scarcity will be the 21$^{st}$ Century's biggest issue," he says.

Tom, who is President of Agrisystems International, has helped us immeasurably through the years with our organic program. He sees a number of key reasons why tobacco growers should turn to organic farming.

"First, you are producing an enhanced value, fair-market priced crop," he says. "Second, farming can make an increasingly important contribution to the whole growing 'carbon-neutral' issue. Third, by growing organically, farmers are building organic matter and nitrogen reserves in the soil."

Tom, who has been appointed by the President to serve on the U.S. Trade and Environment Policy Advisory Committee, offers a fourth reason why organic farming may have a huge beneficial impact on our world now and in coming years. It may be the most important reason yet: local farmers, serving local and regional communities sustainably.

Sustainability? The word conjures up many images. For our readers, we share this definition from the National Sustainable Agriculture Information Service:

> Sustainable agriculture is one that produces abundant (products) without depleting the earth's resources or polluting its environment. It is agriculture that follows

the principles of nature to develop systems for raising crops and livestock that are, like nature, self-sustaining. Sustainable agriculture is also the agriculture of social values, one whose success is indistinguishable from vibrant rural communities, rich lives for families on the farms, and wholesome food for everyone.

National Sustainable Agriculture Information Service (see Chapter 9, Resources for the Organic Tobacco Grower)

Rows of sunflowers are a telltale sign that an organic tobacco farm is nearby.

Tom, our program consultant, offers an example of sustainable farming in practice:

"For every one percent of organic matter turned back into the soil through organic growing, you are increasing the water-holding capacity of the soil by 40 percent," he says. "Putting organic matter back into the soil is like putting a sponge in the ground—a water holding sponge whose life- and farm-sustaining properties can be drawn from in the future."

We should note that Tom, when he's not traveling around the world on organic farming issues, works his own 200-acre farm in Pennsylvania where he grows organic vegetables, small fruits, tree fruits and hay.

## In the Beginning

Long before the advent of petrochemicals, pesticides and "modern" farming in the middle of the twentieth century, all tobacco was grown the natural traditional way *organically*.

First American Indians, then later European settlers, grew tobacco. Natural insect and pest controls were used, and newcomers to the New World learned from Indians how natural fertilizers and crop rotation would sustain the land. Tobacco was America's first cash crop—and it was grown naturally. It sustained the early settlers, the fledgling colonies and a young nation, including our first president and pioneering farmer, George Washington.

Today, near the end of the 21$^{st}$ Century's first decade, societies around the globe are demonstrating greater understanding of the importance of sustainability. And consumers are increasingly demanding that the people who produce the products they purchase and use rely on earth-friendly, traditional and innovative practices.

It wasn't always so.

American Indians were the first to plant and cultivate tobacco—and in an earth-friendly way.

By 1989, some two decades ago, Santa Fe Natural Tobacco Company had gained a small cadre of loyal customers. Founded seven years earlier (1982) in Santa Fe, New Mexico, on the principle that tobacco should be enjoyed the way American Indians intended it, we produced 100 percent additive-free tobacco, natural and with no flavorings or artificial ingredients. As the company grew, we were constantly looking for ways to honor and respect the land from which our products come and the people who help produce them. The search and work toward organic was a natural progression.

The birth of the organic tobacco market is a colorful story. (We try to do justice to that story later in these pages.) Essentially, we worked with an eclectic group of people who shared a common desire, including a pioneering (and a bit eccentric) local grower deep in the heart of tobacco country in North Carolina, a key agricultural scientist from North Carolina State University and an American Indian grower. All were intent in growing tobacco the natural—organic—traditional way.

Along the tobacco road to organic farming came a first step that turned into a successful earth-friendly program. We pioneered the first reduced usage of pesticides program—later to be called purity residue clean, or PRC for short—in the industry.

In the years that followed, plenty of work—much of it through trial and error—went into putting the organic program together. After a number of fits and starts, two growers, who already had some of their farmland laying fallow for three years, began growing tobacco a different way.

Within a few years of setting out, the first small amount of organic tobacco made its way to market. The year was 1991.

Today, more than 100 farmers provide us with organic leaf. That's some 40 growers in the United States, another 40 in Brazil and some 20 or so in Canada. We are also considering working with growers in Argentina and Turkey. In size, the farms run the gamut—from large to small in acreage. These growers represent almost the entire world production of organic tobacco. While these numbers may seem small to many, especially to the big tobacco companies, demand is growing significantly. Organic tobacco production has doubled in each of the last few years.

For their commitment and hard work, organic growers are receiving up to two-and-a-half-times the money per pound for organic tobacco versus conventional tobacco.

Our growers are heartened by this new and profitable market. Many are expanding into other organic products. They are worrying a lot less about petrochemicals—the cost and the risk of mishandling of them. And many tell us they are seeing a return of long-missed wildlife and nature to their land—family farm land that they may well be able to hand down to their children and grandchildren.

## Then and Now: the Old, the Modern, Now the Post-Modern

Not so long ago, tobacco farming was the mainstay of small family farms in North Carolina, Virginia and many other tobacco-growing states. Most fields were five acres or less and carefully tended by a small army of family and neighbors. Whole communities shared plant beds and helped transplant each other's fields. When summer vacation began, kids took over the duties of keeping the growing stalks free of sucker-shoots and hornworms, often because expensive chemicals were not affordable.

After curing was complete, everyone helped load the truck and then followed the precious cargo into town and the tobacco market.

More than a few prayers were said as the auction line moved down the floor and buyers called out bids. If the quality was good, tobacco might fetch an extra 10 cents a pound—that would pay for new farm equipment or a kitchen appliance. The Dixie family farm might have raised corn and cattle, truck crops and chickens, but tobacco paid for household extras and put clothes on the children to go back to school in the fall.

Farming families were the mainstay of tobacco growing. That tradition continues with organic farming today.

Much changed over the years. Modern agricultural practices called for maximizing yield with chemicals and pesticides. Many tasks became mechanized. The federal buyout of tobacco farmers came and the allotment system was gone. The centuries-old auction system was relegated to history, and many farm families stopped growing tobacco. A lot of farmland was consolidated, and many family farms simply stopped growing tobacco or were forced to sell out to suburban developers for the growing migration of people to the south.

Along the way, more people began buying organic products produced through earth-friendly practices—including tobacco. And some of the early consumers and producers were stereotyped.

Today, when you look for the organic tobacco patch, don't look for a farmer wearing beads, driving a VW bus or wearing a tie-died shirt. Look for the sunflowers—the telltale sign of pioneering farmers growing the golden leaf free of prohibited chemicals, fertilizers and pesticides. Instead of ammonium nitrate, the growers use chicken manure and bone meal for fertilizer. Instead of chemicals to control the "suckers" that sprout from tobacco stalks, they use vegetable oil, as farmers did 100 years ago. And they plant a buffer of the tall yellow sunflowers around their organic leaf.

"That's the whole point," R. Lane Mize told us as his seven-year old son, Robert, swatted at the bees hovering near sunflowers that sprout around the perimeter of the tobacco he grows for our company. The flowering plants, he explains, attract ladybugs, bees and stilt bugs that eat the eggs of aphids and hornworms that can plague the tobacco crop.

Lane grows about 60 acres of tobacco by conventional means and seven acres organically at his farm near Oxford, North Carolina.

"I originally had doubts," he said, "but eventually approached Santa Fe Natural Tobacco Company ("SFNTC" from here on) because of the prices the company pays for the organic leaf. I was a little skeptical, but the premium brought me to 'em."

Ball family members "sucker" the plant. Suckers are branches that a tobacco plant likes to sprout, taking energy away from the leaves that the farmer wants to harvest. In older, traditional farming—or today in an organic field in this photo—the farm family drips cooking oil on the sucker branches. It has the same effect as the chemical used in conventional growing, but must be done by hand, plant by plant, because the oil does not easily go through a sprayer.

Lane was one of the first organic farmers to cultivate tobacco for SFNTC. He liked the nostalgia of raising tobacco the way his family did before the advent of fertilizers and chemicals, he remembers. "It reminded me of how we did it when I was young. Plus, that was extra money coming in."

Many of the first organic growers were close to the company's manufacturing operations in Oxford, Granville County, North Carolina, where they sought help in understanding and following the strict requirements of cultivating organic tobacco under state and federal guidelines.

Raising organic tobacco today is labor intensive, as it was in the past, but out-of-pocket expenses are lower. And, while the yield is less, about 1,900 pounds per acre compared to the 2,000 to 2,500 pounds per acre with fertilizers and chemicals, the company helps compensate the farmers by offering prices double and up to two and one-half times over conventional tobacco for maintaining good sustainable practices.

Because growing organic tobacco requires hand labor, it was in the beginning cultivated on a small scale, sometimes with just a few acres dedicated to it. But in recent years, a number of larger-scale U.S.farmers are planting multiple acres; one has 44 acres planted.

How involved is SFNTC in the process? The authors reply individually.

**Fielding Daniel:** "We have a direct rapport with the growers; they want to cultivate tobacco for us. Working with growers on an agronomic basis, SFNTC provides a one-on-one relationship. A number of outside experts have also lent valuable and expert support, such as Peter Hight and David Dycus, both North Carolina Department of Agriculture Agronomists, as well as North Carolina State University experts, like retired entomologist Dr. Sterling Southern, who has helped our organic and PRC growers to find solutions to nutritional and insect problems in the field.

Organic grower Randy Ball, left, received expert help from North Carolina Department of Agriculture researcher Peter Hight, right, and assistant commissioner Dr. Richard Reich, center.

"Most conventional tobacco growers use a certain amount of chemicals no matter what. If they see one bug or 100 bugs—they do anything to get rid of them. On the other side, our growers have been well educated on conservation, environmental issues and chemical analysis issues. These growers are a step ahead of all the other growers that are out there."

**Mike Little:** "When we first tried to set up an organic program back in 1989, farmers laughed at the idea of growing organic tobacco. Now they see it as a way to get a fair price for their crop—double the price paid for conventionally grown tobacco—and we are encouraging them to grow other organic crops, like grains, sweet potatoes and other vegetables."

**Fielding:** "We sample and test every bale of organic tobacco. We check all the way through three pages of banned chemicals, testing each bale of the organically grown tobacco for residues of banned chemicals. We take this very seriously. Thankfully, we've only had to reject two bales of tobacco over all the years of our organic growing program. That shows the dedication of our farmers to organic principles.

"Our contract grower agreements give farmers peace of mind. They know that they can sell their crop at harvest time. And they know they'll get a premium if they meet our quality standards. Actually, calling it a 'premium' may not be quite accurate. We do pay twice as much, and up to two-and-a-half times more for organic tobacco leaf than for conventional. But, as Tom Harding, our organic consultant and a leader in organic circles here and abroad points out, our growers produce an enhanced values, fair-market crop. By practicing and using good sustainable farming techniques, organic farmers are not shipping prohibited pesticides and chemicals downstream for others to bear such costs. The higher prices they receive are earned."

SFNTC's Mike Little

**Mike:** "We're committed to organic and will continue to promote organic tobacco production. In fact, we plan to start transitioning our earth-friendly PRC growers, who already avoid using most synthetic chemicals, fertilizers and pesticides, to certified organic production. Although organic farming is more labor-intensive and requires land to lie fallow for three years before certification, we are committed to its principles in order to continue using the best possible tobacco in our products. Sustainable agriculture promotes the interests of small, independent farmers, not only for tobacco, but also for the organic vegetables and other crops grown in rotation with it. More organic production is in line with our company's principles and better for the environment."

SFNTC's Fielding Daniel

**Fielding:** The price we pay is warranted. We feel like our organic growers are producing crops that were under more stress and a whole lot more susceptible to other things, so we wanted to pay them a little more. The prices are not guaranteed. Growers are paid only if their tobacco tests free of prohibited pesticides. There's a near-zero tolerance under the USDA certification rules (must not exceed five percent of allowable residue tolerance). If we find something that exceeds the rules, then it won't make organic. We have to get away from it and leave it alone, not that we won't take that tobacco, but it loses organic certification right there. We will use it in a conventional blend or something along that way. The biggest thing we want them to do is follow those USDA guidelines. Everybody's got to follow those same rules in order to get that certification."

**Mike:** "We work hard to establish and maintain good relationships with all of our growers. We are in communication with them year-round, and I am proud to say we never lose growers. One reason is our high level of remuneration over the conventional price, based on the grade received. We have an attractive contract. Growers understand and like that about us.

"The geography of organic tobacco has largely been in the 'Old Belt' of North Carolina and the Virginia piedmont, Kentucky and a little in the coastal plain of North Carolina as far down east as Wilson. Now we are going farther south to areas like Lumberton and Fayetteville and east to Kinston and Goldsboro. We are getting farther into Virginia and also may soon move further into South Carolina."

**Fielding:** "Organic tobacco works for two types of farmers: small growers with about 15 acres of tobacco, two or three barns, and enough family labor that they don't have to hire anyone, and enough land to rotate and maybe get some of their income from an off-farm job. On the other hand, organic tobacco fits well for the 'mega' grower, who is already organic in something else so that he can work tobacco into the rotation.

"We believe that certified organic tobacco presents a unique opportunity for tobacco producers. Growers driven to rent their acreage to other farmers due to high chemical and equipment costs may again be able to grow 5,000 to 10,000 pounds of tobacco and make a greater profit. Elimination of prohibited chemicals makes the tobacco field and surrounding areas environmentally friendly, too."

**Mike:** "Organic certification also allows the growth of other high-value seasonal crops, which can demand a premium price on the ever-expanding organic market. Putting five acres of tobacco into organic production makes more efficient use of labor, as the extra work suckering organic takes up some of the lag-time between topping and pulling the tobacco. While some extra labor is required to grow organic tobacco, the increase may well be more than offset by the premium price paid by SFNTC, and by lower chemical costs."

# History of Organic Tobacco in America

The story of organic tobacco growing in America is as colorful as the history of Santa Fe Natural Tobacco Company itself. Fortunately, in researching that history, we had the help of numerous people familiar with the company and its efforts to bring organic tobacco to market.

In many ways, the move to organically grown tobacco was a natural progression. The company has always been committed to providing the most natural tobacco products possible. From the outset, that meant making sure nothing was added to the tobacco from processing and manufacturing through to finished product.

It also dawned on the company that if it was taking every opportunity to ensure the most natural product possible—why limit that to just the manufacturing process? Why not start at the beginning with organic seeds and organically grown tobacco?

It made sense, especially to those in Santa Fe, New Mexico, one of a number of forward-thinking places in America in the 1980s where the organic movement was gaining a foothold and growing, albeit slowly.

So the company looked in its own backyard. What could be better than home grown? Ingeniously, SFNTC worked with two of the eight northern Pueblos of New Mexico—the nearby Tesuque and San Juan (now known as Ohkay Owingeh, its original Pueblo name). Though much of New Mexico is a desert, both Pueblos had the nearby Rio Grande from which to draw water, and they were going to be paid a premium far above what conventional growers in the U.S. Southeast were receiving.

It was a noble effort.

But there is a reason why America's Southeast—places like North Carolina, Virginia, Kentucky and other nearby states—has a reputation for growing the best tobacco in the world. The growing season climate is hot, receives a lot of rainfall and is exceptionally humid, conditions that allow tobacco to develop extremely well. The plants hardly got in the ground in New Mexico's high desert when it became apparent that it just wouldn't work. Added to this were some complicating issues revolving around the official U.S. tobacco allotment system, still in place at the time, that dictated who could grow and sell tobacco, and where.

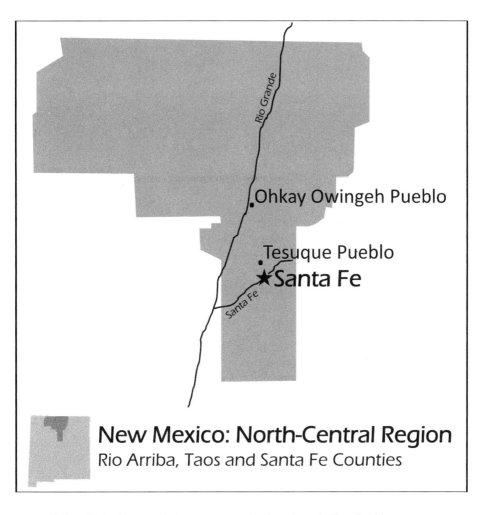

**New Mexico: North-Central Region**
Rio Arriba, Taos and Santa Fe Counties

Early efforts to grow tobacco on nearby American Indian Pueblos near Santa Fe in northern New Mexico.

EACH DOT REPRESENTS 5,000 ACRES

Most tobacco is growing in the Southeast where rainfall is plentiful.

Undeterred, and now in the late 1980s, SFNTC's leaders travelled over the Southeast region's fabled tobacco roads searching for an opportunity and some like-minded individuals who would be willing to grow tobacco differently. It wasn't easy. While organic growing—mostly vegetables and other edible produce—was gaining a foothold out West and in some pockets of New England, it was slow to arrive in North Carolina. Less so today, in the late 1980s North Carolina was conservative politically, socially and agriculturally. Organic tobacco was unheard of. No one was growing it; no one except one individual who, if he were alive today, would describe himself as a bit of a renegade: Brownie Van Dorp.

Brownie's parents escaped the Nazi's march into Holland in 1940, moving to America and finding their way to eastern North Carolina. On the shores of the Pamlico River, near Albemarle Sound and the small town of "Little" Washington, Brownie grew up on the family farm. Not so surprising for a family from Holland, the Van Dorps grew flowers, especially peonies, and vegetables. Later, Brownie experimented with a lot of different herbs and tobacco, growing it naturally, likely not following any prescribed growing regime, but probably as close to organically as we know it today. It wasn't clear if he even had an official tobacco allotment allowed under the system in place.

SFNTC's leaders at the time, Robin Sommers and Leigh Park, were consulting with Southeast agronomists and other experts and academics from such places as North Carolina State University.

They came in contact with Micou Brown, an entomologist, and Albert "Sun" Butler, a chemist and an American Indian tobacco grower. Another early participant was Steve Upton, a grower in southern Virginia.

Photo circa 1989. Among those instrumental in early organic efforts were grower and chemist Albert Sun Butler, left, SFNTC's Robin Sommers, center, Virginia grower Steve Upton, right, entomologist Micou Brown (not pictured) and Leigh Park (also not pictured).

Brownie Van Dorp demonstrated that organic tobacco could be grown. Robin and Leigh consulted with him and reached agreements with Micou and Sun to grow tobacco organically. But Robin, Leigh and the others knew they would have to get quite a few more growers on board to make a go at creating a bona fide organic tobacco product. Robin met with many organic groups and organizations around the country. Robin and Leigh reached out to many growers, but not many responded. The few who did included respected conventional flue-cured tobacco grower Ben Williamson, from Darlington, South Carolina, and burley grower Roger Smith, from Brooksville, Kentucky.

Photo circa 1989. Meeting with grower Ben Williamson—one of the first organic tobacco growers—far right, were Micou Brown, left, a photographer, and Albert Sun Butler, next to Ben.

Most growers were not so quick to take a chance on a different way of growing. "Who were these people from out West talking about *organic tobacco*," one could hear the farmers say.

Meanwhile, as this early effort to create a market for organic tobacco was in its infancy, many of the parties involved debated how to proceed. Some wanted to forge ahead and immediately produce organic tobacco products, primarily because they thought it was good for the environment. Others were concerned about setting up procedures and following prescribed regulatory guidelines that would allow the products to carry an organic label. The infrastructure to get tobacco certified did not even exist. SFNTC decided that the organic program would evolve deliberately, even if slowly. However difficult it was, the effort to recruit growers to organic would continue.

But the company did decide to move forward aggressively with a program to enlist farmers to not use certain systemic chemicals that leave residues on the tobacco. This program was later to be called PRC—purity residue clean.

Earth friendly PRC—purity residue clean program—was introduced first while organic infrastructure developed.

"Our thinking behind this," says Leigh Park, "was that we (SFNTC) had long offered the most natural tobacco products we could. We eliminated the manufacturing additives, the non-tobacco items in the manufacturing process. But then we realized there was the issue of agricultural residues in conventional tobacco growing. So we set out to reduce that." Hence the SFNTC-PRC plan was born.

The PRC program grew rapidly. It wasn't that difficult to encourage farmers to limit the use of a number of systemic pesticides, but the organic program—which required a much greater change by farmers—moved forward by inches. However slow it was, progress toward organic growing was a move in the right direction.

It wasn't until a man who hailed from the heart of tobacco country joined the effort that the organic program made real headway.

Today, Michael Little is senior vice president of operations for SFNTC. He's known as "Mike" by nearly everyone, from the home office in Santa Fe, where he spends just a few days each month, to the many organic tobacco growers he visits in the Southeast, to coworkers in the company's manufacturing operations in Oxford, North Carolina.

At about the age when kids first walk onto a baseball field, Mike was already in the fields—tobacco fields—around his Saratoga, North Carolina, home learning and working. Mike's hometown, near Wilson, in Wilson County, North Carolina, is deep in the heart of the Tarheel state's "bright belt," flue-cured tobacco region. He reflects the area's proud tobacco heritage.

"Tobacco," he says in a respectful, soft-spoken manner, "is one of America's original crops.

"Our region of the country has a deep heritage, having funded the American Revolution and making it possible for one of the world's first democracies to succeed," he says almost reverently. "It's one of the reasons why visitors to Washington, D.C., can see image after image of tobacco leaf in paintings and architectural relief adorning the Capitol building and many other structures in the nation's capital. Tobacco remains an important part of our economy, but it's also a big part of America's history."

Mike Little, head of SFNTC operations and master blender, has been working with tobacco nearly his whole life.

America's first cash crop, tobacco, supported a fledgling democracy.

Mike is a master of the art of tobacco blending—the demanding skill of selecting the right styles of tobacco from different harvests to achieve just the right taste. He credits the years spent learning at the hands of an industry master, a gentleman by the name of Jack Smith III, a Virginian. "He was a tobacco blender's blender," Mike says. "He knew tobacco leaf like the back of his hand and had a real feel for it—less of the modern R&D and financial analyses of the big tobacco companies today, and more the handmade art of blending, knowing and caring for the various leaf and its many characteristics."

While Mike did not officially join SFNTC until the mid-1990s, his ties to the company began a decade earlier. He was working with an independent tobacco operation in Petersburg, Virginia, that supplied tobacco to a small New York City company making handmade cigarettes for some of the nation's wealthiest families and consumers. As real estate prices in the financial district in and around Wall Street skyrocketed, the land beneath the small operation became too valuable for the owners, who were also getting older, to continue the operation. They turned to Mike and his partners, suggesting that Mike deal directly with the firm's high-end customers who wanted only natural products of the highest quality.

SFNTC was at the time purchasing tobacco from the New York City firm and three other contract manufacturers. Mike was introduced to the then-head of SFNTC, Robin Sommers, who quickly saw in the master tobacco blender a man who could be trusted to serve SFNTC's growing number of consumers. A decade later, SFNTC, with its business growing double digits each year, decided it needed to form its own manufacturing operations. The company turned to Mike again and in 1995 asked him to join SFNTC and create and manage the company's manufacturing operations in Oxford and its fledgling organic program.

With Mike at the helm of tobacco operations, growers interested in trying this new old way of growing tobacco began signing up. Many of the early organic growers came from the PRC program. "Reducing the use of pesticides and the residue it leaves was a great idea," one grower told Mike. "It worked fine for me, so why not try the whole shootin' match and go organic."

Helping make the program move forward in a big way was the addition of Fielding Daniel in year 2000. Fielding, now director of leaf and blending for the company, is an experienced leaf expert. He works closely with the growers and is greatly respected.

Fielding Daniel, SFNTC's head of leaf, works closely with growers.

As the program developed, it became apparent that organic tobacco not only required more labor on both the part of the grower and the manufacturer, it also required a rigorous approach to documenting, certifying and reporting. To help the company navigate these unplowed fields of government regulations, SFNTC turned to one of the most experienced and respected experts in the field of organics, Thomas B. Harding. Tom is a founding member and past president of the Organic Trade Association and current board member of other world organic groups.

Tom Harding, left, a founding member and past president of the Organic Trade Association and current board member of other world organic groups, provides expert advice to SFNTC and its organic growers. SFNTC's leaf manager Randal Ball, center, and leaf director Fielding Daniel, right.

He was, and continues to be, instrumental in creating and keeping SFNTC's organic "Bible," the company's Organic System Plan, and maintaining organic integrity throughout the entire agricultural system, from farmer to consumer. Tom first started consulting with SFNTC in 2002 and continues to this day as an instrumental partner in the effort.

Two decades after first exploring and experimenting— and after many bumps along the way—more than 100 farmers are growing organic tobacco for SFNTC. In 1989, the company bought and processed 4,000 pounds of organic tobacco, which made its way to market after being aged for two years. By year 2000, some 60 growers produced 750,000 pounds of organic tobacco for the company. In recent years, demand for organic tobacco has been doubling annually. In 2008, SFNTC processed more than 2,000,000 pounds.

*3*

# "In Their Own Words"

C ollected from various sources and over a number of years, the actual words of many of our organic tobacco growers appear on the following pages. Where we've added some background and context, we've placed our words in italics.

## Flue-Cured Organic Growers

### Ben Williamson, Darlington, South Carolina

*One of the original growers contracted by SFNTC to grow certified organic tobacco, Ben Williamson started growing organic tobacco in 1986 and raised it through the 2000 crop year, after which he retired. He started by planting one acre his first year and increased his acreage each year. He had 12 acres planted his final year before retiring.*

*We've always had tremendous respect for Ben as a grower and as a person who sets out to not only become an expert at what he does, but a real steward of the environment.*

"For me, organic tobacco is more profitable than conventional tobacco, despite the organically grown crop's high labor requirements. There are a lot of things I like about this crop.

Ben Williamson, whose farm is pictured above, is a true steward of the environment, and was one of the first organic growers. He subscribed to many sustainable growing practices, such as planting wheat between rows of organic tobacco to provide a wind break to hold the soil and other plantings to attract beneficial insects.

"I feel that organic growing has environmental benefits, too."

*Historically, tobacco farmers have been among the heaviest users of crop chemicals, including substances like MH30, which was often used to remove suckers from the plants under conventional growing.*

"When farms moved over to organic production, all the sucker control had to be done by hand, and that was probably the biggest job.

"There are fungi that spoil the tobacco, and some viruses, and then it's a heavy feeder so a lot of fertilizer is required. But sucker control is probably the biggest reason for the use of chemicals. And, we even used some pretty powerful systemic insecticides, which I'm sure remain in the soil.

"I find in SFNTC the most agreeable and energizing approach to farmers I've ever seen. They are a friend of the farmer and are very forward-looking and progressive."

**Billy Carter**, Eagle Springs, North Carolina

*A young farmer who grows more than 200 acres of purity residue clean (PRC) tobacco for SFNTC on his Moore and Montgomery county farms, Billy first grew two acres of organic tobacco in 1998. He now grows organic tobacco on 42 acres. He is the largest single organic grower for SFNTC. As to growing the crop, Billy says:*

"I like the challenge of it. Though organic leaf requires more labor, it's not that it's so much more difficult to grow. The biggest thing that (SFNTC) pays more for is the increased risk of crop loss. Growing organic tobacco is basically the same as growing other tobacco, except for fertility and pest control."

*Several years ago and soon after he'd started growing organic tobacco, the Charlotte Observer wrote a feature story on Billy. He told the reporter:*

Billy Carter

"Hoping that I could grow some specialty-type tobaccos, I was excited when I started growing organic tobacco for SFNTC. Compared to growing conventional tobacco (with chemicals and commercial fertilizers), with organic tobacco, you have a lot more limited options—and then they are more long term and not quick fix options. It's a management thing—being more in sync with what Mother Nature brings you in a particular season. Labor is another critical factor: we harvest our organic tobacco manually, but use mechanical harvesting equipment to bring in our conventional crop. So, there's a much greater level of labor required in growing an organic crop of tobacco.

"Our first crop of flue-cured grown organically was in 1998. We planted about 2.5 acres to achieve a 4,000-pound contract. Over the years, we've tried to do a good job and, in 2005, SFNTC contracted with me to grow 14 acres. From that I sold 31,000 pounds of organic flue-cured tobacco to Santa Fe. We had a pretty good yield in our organic flue-cured tobacco that year. As a result of having delivered good quality tobacco each year, SFNTC contracted us to also grow 2.5 acres of organic burley tobacco in 2006. We found it challenging in a lot of ways, considering that we've never grown the crop, nor even seen it grown anywhere. But, it worked out fairly well.

"We now have nearly 100 acres of certified organic land. We've had some success growing other organic crops, but haven't found quite the right niche for marketing them. We wanted to sell more at the wholesale level, but that's not where the premium is. It's better to market at a local level yourself. We've availed ourselves of good, long-term (three to four year) rotations for our tobacco, which helps us improve the quality of our land in organic tobacco.

"There are a lot of things we enjoy in our relationship with SFNTC, including the fact that they are still a young and growing company and are excited about what they are doing. Obviously, the thing that drives our interest is that there's a premium paid for the product they receive. We feel like it is a well-earned premium—a mutually beneficial-type economical relationship. We put a lot of effort into what we do to go through the certification process and grow the crop in a manner prescribed by Santa Fe. So, we feel they get the value from what they pay us, but obviously what they pay us drives our interest in maintaining our organic tobacco operation.

"Another thing we like about Santa Fe is that there's a much more personal touch involved. They are very interested in what you are doing. It's nice to be around people who are excited about what they are doing—trying to find new and better ways to market their product. They've been very supportive of their grower base in the process. It's hard to heap enough accolades without sounding that you're being overly flattering.

"We sure are excited about the opportunity to grow more tobacco organically and are excited about the prospect that they are developing other markets for their organic product. That would be very beneficial to us on this farm and allow us to pass on this legacy to our children—something we'd find exciting to happen.

"There are a lot of things I like about growing organic tobacco, but what I like most is that the 12 acres I grow pay quite well. The higher price that Santa Fe offers compensates us for the added labor, materials and risk a farmer takes on when I cast away the chemical arsenal that we used to keep weeds, insects and diseases out of our crop. Instead of using pesticides, we plant sunflowers in tobacco fields to draw ladybugs. These eat aphids, which like to suck the life out of tobacco."

*Billy admits that some years, his tobacco fields—filled with weeds and suckers—were some of the ugliest, nastiest looking fields in the state. But, they were also some of the most valuable.*

**William F. Wyatt**, Mt. Airy, Pennsylvania County, Virginia

*A southern Virginia organic and PRC grower, Bill is one of the company's original contracting farmers.*

"Growing organic tobacco has not been easy. But the satisfaction of producing 'pure' tobacco is hard to beat. It's basically what the Indians or early colonists would have done. You have to use only organic materials, doing it pretty much the way they did 150 years ago.

"I've been growing organic tobacco for SFNTC now about nine years. I've really enjoyed growing tobacco for them and think it's a good product. I really like to know I'm growing something that's in tune with the environment and it really makes me feel good doing it that way. But it does take a while to learn and figure out how to grow it that way. When growing organic tobacco, you're looking at doubling your labor and that's a big thing. You're kind of growing tobacco like they did 150 years ago, but I enjoy doing it. You can come out here anytime and see deer, turkeys and all types of wildlife. It's a good thing for the environment.

"Santa Fe makes me feel like I'm part of a family. We have many meetings together and it's almost like a family operation. It's worked real well for us. After we figured it out, it's done real well for us. At this point, I wouldn't be growing tobacco at all if it weren't for SFNTC. I hope that it'll be there in the future so my sons can farm and raise tobacco."

Bill Wyatt

**Glen Preddy**, Wilton, Granville County, North Carolina

"My brother, Jeff, and I have been farming together for 20 years and growing five acres of organic tobacco under contract with SFNTC since 1999. Each year is a new experience, but we're learning something new every year. We've learned how to cope with the challenges of getting the fertilizer rates and pest control correct. And growing the crop organically requires a lot more labor. SFNTC helped us. Their field reps come out and help us, showing us different ways to do things. Overall it's been a good partnership, and yes, I'd like to be growing more of it. I think we can handle it now—after years of experience we've learned a lot."

**Ronnie Moore**, Kenbridge, Lunenburg County, Virginia

"Growing tobacco organically is basically like my grandfather used to raise it when I was a little kid. We're using organic chicken manure and use mineral oil for the suckers. Our crop sometimes looks terrible from the street, but there are no suckers when you are harvesting. We raise about 30 acres of tobacco altogether; about 11 to 12 acres of it is organic tobacco.

"I've had a great relationship with SFNTC. Fielding and Willy Brooks (who helped devise some of the original tobacco grading system for SFNTC and inspects and receives tobacco for the company) have been very cooperative and take into account such things as a dry year and aphid problems. They know that's going to happen with an organic crop. I've been very pleased with the prices they've been giving and where they've been increasing my contract poundage. Any equipment you use outside of the organic fields, you need to power wash all equipment so no soil goes from one field to another. It's just a pure product—basically what the Indians or early colonists would have done. Quail hunters will ask me if I've seen any quail around here. I say they're out in the tobacco fields. We don't use any pesticides or chemicals out there to harm them. That's where they're raising their young— right in our tobacco fields."

Ronnie Moore

## Ralph Tuck, Virgilina, Halifax County, Virginia

*Ralph is one of the original SFNTC growers. He raises both organic and PRC tobacco for the company.*

"I'm proud that I'm raising (organic) tobacco the natural way—without prohibited materials and with approved products. We have improved the amount that we're growing a little bit. We could farm more than we're growing. It's definitely different from growing your regular tobacco. You use the same basic knowledge, but you use different inputs and cure and handle it differently. It takes a lot of time and it's definitely a challenge. It's been very good for us. We're definitely capable of growing more."

Ralph Tuck

**Terry & Cecil Allen**, Roxboro, Person County, North Carolina

"We started growing organic tobacco for SFNTC in 1998 and have been growing it every year since. We've got about 35 acres certified, rotating no less than every three years. We learned how to control aphids, cause you might as well quit trying to kill them. The yield is not that much different from conventional— very similar.

"The biggest difference is the labor and that requires more land for the rotation. Santa Fe has been very helpful in getting us started. They always greet us with a smile, that means a lot to me. We have plenty of land, so we could grow more."

Cecil Allen

## Lane Mize, Granville County, North Carolina

*Lane was another of the first group of farmers to grow organic tobacco for SFNTC.*

"I enjoy growing it. I was a little skeptical of the experience at the beginning, but after I got into it, I learned that it's just basic rules you follow to become certified. I like it better than conventional tobacco. Our organic yielded just as well as our conventional tobacco did.

Lane Mize

"You've got all these guys *(consumers)* from the 1960s who've got money now. I think it's really going to roll. I really think they (SFNTC) are going to expand, and I want to expand with 'em. They seem to treat everybody like a member of the family. Every time I needed anything or called, they were always there. As far as organic tobacco, I want more. I'd love to increase my acreage."

**Jane Iseley**, Burlington, Alamance County, North Carolina

*Jane Iseley was among the first group of organic growers. Jane was working to capture things in their pure and natural state as a professional photographer long before she became a SFNTC contract grower in 1998. Raised on her family farm, Jane became an accomplished photographer but eventually returned to her Alamance County farm. In 1980, her ailing father helped Jane learn to grow tobacco.*

"Growing organic tobacco under contract to SFNTC has allowed me to stay in business. We never have been big tobacco farmers. But, it just sort of fit our operation. If we weren't growing organic, we wouldn't be growing tobacco at all. Growing organic tobacco free of any prohibited pesticides convinced me to cut back on using pesticides on my conventional produce as well. I started experimenting with raising organic produce, such as corn and lettuce. And now I've gone totally organic.

"To think that it all began when I started growing organic

tobacco for SFNTC back in 1998. I had asked my Daddy earlier, 'do you think you could teach me how to grow tobacco?' He kind of perked up and said, 'Yeah, I think I could do that.' So we started with one acre and that first year he had to teach me everything about growing tobacco. We'd leave out every ninth row. In those rows, we'd grow beneficial plants, which attract other insects that can get on our tobacco. We farm basically for the quality of life. It's a hard life. But being on the farm, and being on the river bottom, we offer a lot of ground for animals to raise their young."

Jane Iseley

**Allen and Randy Ball**, Henderson, North Carolina

*These two brothers who grew organic tobacco for SFNTC from 1997 to 2002 are an excellent example of part-time farmers who made a great contribution to moving organic tobacco forward. Great sadness came upon our entire tobacco family when Allen unexpectedly passed away in 2002. But both Allen and Randy's work and dedication continue to be remembered. Randy's sons (Allen's nephews) Ryan and Randal both work for SFNTC in Oxford—Ryan as director of distribution, Randal as leaf manager.*

*For Allen and Randy Ball, part-time organic tobacco growing was a family affair. The brothers grew about four acres of tobacco, with all members of their families taking part, mostly working evenings after their regular jobs, and on Saturdays. Because organic tobacco production is so labor intensive, all of the Ball family members, including wives and children, were involved. Even the Balls' mother, Eva Pearl, helped cure tobacco. The brothers were proud of their farming, growing sunflowers with the tobacco to attract ladybugs that prey on aphids. Randy, for an article in SFNTC's* Smoke Signals *newsletter commented:*

"We may not have the prettiest tobacco, but I'm sure we will have the prettiest fields.

"In farming and growing organic tobacco, you need to want to take care of it. By using organic programs, you get away from using so many chemicals and go back to the way it used to be. And that's what we like about it. With the right rotations, your yields can be just as good as they are with conventional tobacco and you certainly have just as good a quality.

"We've got conventional conservation practices in, like waterways and so forth. We've also been experimenting with no-till tobacco. We're excited about that and plan to continue it, even organically. Santa Fe has been real good to work with. They've allowed us an opportunity to work a small-scale farm where we probably couldn't otherwise with conventional tobacco because of the size of our farm.

"Since both of us are part time farmers, growing just four acres of organic tobacco, all the members of our families take part in working in the evenings after their regular jobs and on Saturdays. It's our goal to keep the small family farm going for more generations."

Randy Ball, Mike Little and Allen Ball, left to right

A young Ryan Ball, Allen's nephew and Randy's son, hand harvests organic tobacco. Ryan is now SFNTC's director of distribution.

*Favorable weather conditions may have been the reason, but organic production was surprisingly profitable for the brothers their first year. Aided by great weather and a low infestation of worms, the Ball brothers averaged 2,400 pounds per acre on their four organic acres—a very good yield, better than some conventional tobacco in the area achieved.*

**Richard Ward**, Whiteville, North Carolina

*Although he had farmed conventionally all his life, Richard grew his first organic tobacco in 2000 and, at last count, was growing 35 acres of the organic crop.*

"The first year I grew organic tobacco we were struck by a hailstorm twice in one day. It was late July and the first hailstorm struck at three in the afternoon, the second hit six hours later. It just devastated the crop; lost 60 percent of it. The second year, I planted 10 acres, all the while researching everything on organic farming to increase my knowledge.

Richard Ward

"The reason I first got into organic growing was the economics. I talked to an agricultural extension agent and asked him what I could do to supplement what I was making. Everything in conventional farming was getting smaller and smaller. This ag agent sent some information about organic crops but, as I learned later, a lot of information out there at the time (about 10 years ago) was inaccurate. So, I got the names and contact information for some farmers who were growing organically. We talked three or four times. And I toured other organic growers, one was 65 miles away.

"I was running some cattle at the time, so I had a lot of acreage that had been with no chemicals for five or six years. I also had some land that I had leased; it was too sandy—not enough topsoil for corn and beans. Well, I started farming it and could see the income I could get was better.

"The way conventional farmers think is this: 'got a problem (like a disease or insect infestation), go buy a chemical and treat it. Well, with organic farming, we have other non-chemical tools, like putting certain plants in to attract good insects that eat the bad ones, and natural fertilizers. With organic farming, you learn management skills.

"What keeps me going is the feeling of satisfaction I get when I see the results of all the hard work that goes into growing everything organically. I've always been a hands-on farmer and like to get out there and work in the fields myself."

*Like many organic tobacco growers who must rotate crops to maintain healthy soil, Richard also grows organic produce. In addition to tobacco, he planted sweet potatoes his first year. But disaster struck again when he encountered a certain vegetable broker who was, let's just say, less than honest. As a result, the next year Richard took on those marketing chores, too.*

Richard Ward with some of his organic produce he sells to Whole Foods.

*He continued to attend all the workshops he could find to learn better ways of doing things. He participated in Carolina Farm Stewardship Association events, U.S. Department of Agriculture programs and North Carolina fruit and vegetable expos and conferences picking up all kinds of information here and there.*

*In his second year of organics, he grew sweet corn, corn for grits, hominy and corn meal and cukes, squash and watermelon. But he didn't make much money. "Had to get known in the marketplace; build a reputation," he says. He later grew peas, strawberries, broccoli, Irish potatoes, which he shipped to Boston, honeydew melons, bell peppers, tomatoes and six varieties of squash.*

*Finally, about five years ago, he formed a partnership with some other growers called Eastern Carolina Organics to get his produce to market. "We hired some workers, bought some coolers and trucks," he says. "Done right good with it." Natural food chain Whole Foods is one of his biggest buyers, and he praises their organized approach to dealing with growers like himself. "There's plenty of face to face meetings and opportunity for the growers to share their viewpoints and give input," he says.*

### **Stanley Hughes**, Orange County, North Carolina

*Growing tobacco organically means many hours in the field doing work by hand. When he experienced quota reductions in the old tobacco system in the 1990s, Stanley's already small Pine Knot Farm was made even smaller. (Stanley, who grows and sells organic vegetables, was honored as "young farmer of the year" in the early 2000s by North Carolina A&T University.)*

"I wouldn't have been able to farm if it hadn't been for the organic tobacco opportunity.

"I'm a third-generation African-American farmer, the youngest and the only one of 12 children who pursued farming as a career. For more than 20 years, I had an off-farm full-time job, farming tobacco part-time, growing it organically since 1996. To continue making a profit on a farm whose tobacco allotment was whittled down from 25 acres to 15, I had to turn to organic. I was one of the first farmers in the state to start growing organic tobacco under contract to SFNTC.

"The only real difference in growing organic tobacco was that I had to learn how to grow a crop within the strict organic standards. It's not all that much different than when I was tending tobacco with my father with no chemicals available to control pests. I'm a real believer in organic tobacco, particularly because of the favorable prices SFNTC pays when you bring in a certified organic crop of quality tobacco. By producing quality leaf, using environmentally friendly chemicals and proven cultural practices, I've been receiving a premium price.

"With the high cost of farming, if you're not a larger farmer, you just about need to do other things if you're trying to stay on the farm. Being African American, I felt there wasn't much future in farming—to get loans and get them on time, you had to put up more collateral or have someone co-sign with you.

"I've put my sweat and ingenuity into the same land that was tilled, sown and plowed by my maternal and paternal grandfathers before me. At one time, this farm supported three families and now it's down to where it's supporting one, but I feel good to be the only one carrying out our grandfathers' dreams.

"To be able to be free, work out in this air and to see the things that I can now grow organically, is what it's all about. It's sort of like saying, 'I put this seed in the ground and look what I made of it.' Of course, without the Good Master helping you,

you ain't going to accomplish nothing, no way, but to be able to say you guided this is what farming is. It's taking a path of your own.

"That first year, I designated one acre, and after I planted that one acre, I saw that I should have tried two."

*Today, Stanley is growing six organic acres, along with nine acres of PRC tobacco.*

"The organic method is a little bit more risky because you can't go in there with any kind of prohibited material, but the only thing about the organic is that you're willing to take the chance for something when you think you're going to get twice as much from the crop."

Stanley Hughes

**Tommy Winston**, Granville County, North Carolina

*Tommy Winston was another of the original contract organic growers, and grew the crop successfully for several years.*

"SFNTC folks are mighty nice to work with. When I had trouble finding fertilizers that would qualify as being organic, I thought 'why couldn't we try something used by American Indians here, and use fish meal.'"

Tommy Winston

*He recalled stories about how Native American Indians used to put a fish in the hole with their corn seed. He knew fishmeal was used as a feed supplement and a fertilizer. Tommy thought about using mineral oil as a sucker control method, because he remembered using it when growing tobacco with his father in the days before chemical sucker control agents were developed. He uses either corn or soybean oil as a substitute, because they are easier to get than mineral oil, although there may be some leaf drop if too much is applied. He says he still has to go back and hand sucker though. Tommy contracts with SFNTC because he gets a higher price for the organically grown tobacco, and because he feels the company appears to be willing to make the extra effort to work with growers. He doesn't see a lot of difference between growing tobacco organically or conventionally—but organically grown tobacco does require more chopping and cultivating for weed control.*

## Burley Organic Growers

*Organic burley tobacco growers are fewer in number than their flue-cured counterparts and are found mostly in Kentucky. For those who have tried growing tobacco organically, it has been a good experience.*

**Roger Smith**, Brooksville, Kentucky

*Roger was one of the first farmers to grow organic burley in the early 1990s. Today Roger grows and coordinates seven other Kentucky growers under contract to SFNTC.*

"I personally would not be raising tobacco now unless it was organic. But you have to work hard at this to make it work. A lot of growers have approached me on how to grow organic tobacco. It also helps that Santa Fe is a farmer-friendly company and is out there supporting the farmer.

"Insect control is a challenge, too, for burley tobacco growers. That's because the only commercially available insecticides that can be used are derived from Bacillus thuringiensis (B.t.). We use Dipel to control hornworms that affect our crop. But we've learned to promote populations of beneficial insects like ladybugs and lacewings by planting crops that attract them. They like to breed in sunflowers, hay and certain flowers.

"Most chemical farmers back away from organic tobacco when you tell them they have to sucker by hand. But it can be done using varieties that don't sucker a lot. Burley growers need to sucker their crop twice in most years. But in a wet year, it may take three. Growers get a degree of sucker suppression by applying soybean oil to the tops and letting it run down the stalk.

"The best candidate for organic burley is a farmer who has some bottomland soil. But you have to watch out for a polluted river that might flood and pollute your soil. The same thing can happen with irrigation water and might decertify you as organic. Organic burley will be confined to small-scale growers until there is more demand. But the interest is definitely growing. The price offered by SFNTC is generally about double the market price. Farmers deliver their tobacco to SFNTC at a warehouse in Lexington, Kentucky, on a specified date, usually in December."

Roger Smith

**Gene Turpin**, Lebanon, Kentucky

"There are some definite benefits from growing tobacco organically, especially improving the character of your soils. Of course, you never get something for nothing. Organic tobacco takes a lot of hard labor. You spend less on chemicals, but the workload is increased.

"An added requirement in growing organic tobacco is that you have to find a hoe that fits your hands pretty well—that's because weed control is entirely by hand or mechanical cultivation. Even though a lot of lessons must be learned, I would recommend organic tobacco to any other farmer."

Kentucky burley tobacco.

**Tom Croghan**, Cub Run, Kentucky

*Tom has been growing organic burley tobacco since the late 1990s.*

"My organic yield is usually between 2,000 and 2,200 pounds per acre. Occasionally you might get 2,500 pounds per acre. That compares well to my conventional tobacco, which will yield between 2,800 and 3,000 pounds per acre. Because SFNTC pays a substantial premium for organic tobacco, it's been a real lifesaver for this farm.

"I've often used Safer Soap in a few places for aphids. I don't like to use insecticides because most kill the beneficial as well as harmful pests. You have plenty of ladybugs, lacewings and praying mantises anyway. I also have a pretty good population of parasitic wasps, so we don't see very much damage due to worms."

Staked burley being harvested.

**Chris Korrow**, Burkesville, Kentucky

*Chris successfully melds production of an in-demand organic crop with centuries-old methods. In 2000, he was among eight Kentucky growers and six in other burley states who produced a total of about 70,000 pounds of organic burley for SFNTC—an average of roughly 5,000 pounds per grower.*

"I think organic tobacco has a brighter future than some conventionally grown tobacco because it is part of the growing organic trend in the country."

# Growing Organic Tobacco

*I*n this chapter, we provide information about the process of producing organic tobacco—from preparing the soil and certification, to planting seedlings and bringing cured leaf to our receiving stations. That's a lot of ground to cover and we'll try to be as thorough and clear as possible. But, as we pointed out in our introduction, and as in any pioneering effort, much of what has been done to bring forward a better way of farming has been through trial and error. We don't presume to have or present the last word on organic tobacco farming. But we and the organic tobacco growers have learned plenty over these last two decades. By sharing this with you, we hope we can encourage more growers to farm organically.

## The Certification Process

To gain organic certification, growers must select an area that has been without chemical fertilizer, herbicide, insecticide or other synthetic chemical applications for three years prior to the beginning of the crop year. Land that has been "rested" or fallow and is high in organic matter is often the most suitable. Land that is already certified organic and used for other organic crops may also be suitable.

Individual fields may be certified as long as appropriate buffer zones, such as meadow strips, exist between organic fields and conventionally farmed areas. Consulting agronomists and a certification inspector visit the farm to inspect the designated fields and to review the farmer's crop production methods for organic certification.

North Carolina Department of Agriculture and Consumer Services' (NCDA&CS) agronomist Robin Watson sits with grower Jane Iseley to go over paperwork at her roadside organic produce stand.

The certification process takes about three months and the application today costs about $25. (See Chapter 9—Resources for the Organic Grower—and the section, Quality Certification Services, QCS, for detailed information on the certification process.) Growers are provided a list of approved nutrient sources, as well as weed, insect, sucker and disease-control products and methods.

Organic tobacco falls under the purview of the United States Department of Agriculture (USDA) National Organic Program. (See Chapter 6—USDA.) Both the contract growers and the manufacturing and storage facilities must be certified to claim the organic label. The company runs tests on tobacco in the field as well as at the receiving station to verify it is residue free.

Organic farmers must have their crops certified by an official inspector. In addition, SFNTC tests the tobacco at our receiving stations to make sure that it has no prohibited chemicals on it. To achieve organic certification, tobacco farmers must follow a strict, labor-intensive growing regimen. That requires a lot of detailed record keeping, making sure everything that happens with the crop is recorded accurately and in a manner that gives an inspector a clear picture of what the grower has done to maintain certification.

Since certified organic tobacco is grown without the use of prohibited pesticides and fertilizers prohibited in USDA's National Organic Program, even equipment used outside organic fields must be power washed so that no soil goes from one field to another.

"When we started growing organic tobacco," says Eagle

Springs, North Carolina, grower Billy Carter, "we were fortunate to have a couple of fields that were in long-term, conventional rotations that had no commercial fertilizer or pesticides applied to them. As we tried to get more ground certified organically, we had to be sure there were buffers as required from rights-of-way and other fields and common sources of water for irrigation. You have to be concerned about not contaminating your crop. You've got to be aware of what's going on all around your organic operation."

While all growers agree that organic farming requires additional work, our growers vary in their opinion of the amount of additional labor required and the degree of difficulty in achieving and maintaining organic certification.

"It's a year-round process," Billy says. "I keep a notebook of all the activities I'm involved in with the organic crop. Then I record it in a manual we're required to keep. It takes a lot of my attention, given the small acreage involved. So, you have to be aware of everything going on around your crop to show an organic inspector that you've truly kept up your part of the bargain.

"The fields must be free of prohibited chemicals for at least three years before the crop is planted and must have buffer areas between them and conventionally treated fields," says Billy, who raises about 2,400 to 2,500 pounds of tobacco per acre on his conventional fields and about 2,000 pounds per acre in his organic fields.

When Randy and Allen Ball converted their land to growing organic tobacco, they said the change required no special equipment, and the small operation required only family labor.

Grower William Wyatt found that the separation from conventional crops, the detailed records of planting, cultivation and application of crop dressings required a lot of hand labor.

Of the certification process, Richard Ward, from Whiteville, North Carolina, said the main rule for growing organic foods is "that no prohibited chemicals can come in contact with the crop." Equipment and machinery, he said, "must be power-washed clean of any dirt from conventionally grown crops, and no prohibited pesticides can be used." Even the chickens, which produce the manure, may be fed organically grown corn. "It's the same plant in the end. You just can't use prohibited chemicals."

## Growing Organic Tobacco Seedlings

Quality transplants are important in producing organic tobacco. Organic transplants can be raised in the greenhouse or plant bed. SFNTC agronomists and other outside resources (see Chapter 8—Resources for the Organic Grower) can help growers select the variety that best fits the needs of their specific growing or marketing conditions.

Growers must rely on disease resistant varieties and crop rotation as their best defense, because there are no organic control agents for diseases such as blue-mold, black shank and Granville wilt. Careful management with sanitation, weed control and natural fertilizers helps to grow healthy, disease-free transplants right from the start.

Grower Billy Carter and NCDA&CS agronomist David Dycus, foreground, take a close look at organic tobacco seedlings in greenhouse flats.

## When to Plant the Seedlings

On his South Carolina organic farm, Ben Williamson (now retired) transplanted his seedling plants to the field the first or second week of April. He set out his plants using a one-row setter in a tilled field. By contrast, in the upper Piedmont region of North Carolina, transplanting usually doesn't start until mid-May.

Farm workers transplant tobacco seedlings.

## Where to Get Seedlings

Grower Billy Carter of Eagle Springs, North Carolina, produces his own transplants in his greenhouse. Organic rules require that transplants used in organic production must be grown with natural fertilizers. That's not so easy to come by, so Billy started experimenting with the help of the North Carolina Department of Agriculture & Consumer Services (NCDA&CS).

Tobacco transplants, as well as many vegetable transplants, are grown in float beds, where seedlings sit in Styrofoam trays that float on top of a nutrient solution. While commercially produced fertilizers are ideal for float beds because they dissolve easily in water and do not clog pumps, organic fertilizers do not possess these same properties and using them in a float bed is pretty much uncharted territory. After studying various alternatives, Billy picked two he wanted to try in the float beds: seabird manure and bat guano. That's when he called on NCDA&CS agronomist David Dycus to help him evaluate these fertilizers.

Seabird guano mixed in a solution is used by some organic growers for seedlings as natural fertilizer.

After checking the nutrient solution of both products, calculations showed that it took three times as much bat guano as seabird manure to provide enough nitrogen for the seedlings. The bat guano also cost twice as much as the seabird product. But as the experiment continued, an alkalinity problem surfaced—dangerously high levels of nitrogen resulted in plant roots not being able to grow in the solution without being burned off. He soon found that the plants rooted just fine with the bat guano. Thanks to this solution-analysis project, Billy was able to produce some field-ready organic transplants the next year.

Tobacco seeds, which are very small, are sometimes coated in natural clay for ease of handling.

In growing organic seedlings, growers strive to get well-germinated, uniform plants that have been clipped correctly.

Four good sources for purchasing organic seeds are: Gold Leaf Seed, Richard Seed, Workman Seed and Cross Creek Seed.

Organic grower Jane Iseley gets her seedlings from organic grower John Vollmer of The Vollmer Farm, Burr, North Carolina.

## Preparing Land and Fertilizing for Organic Tobacco

It all begins with the soil. Organic farmers depend on the earth, so the healthier the soil, the better to grow successful crops. It doesn't matter if it is for tobacco or fruits and vegetables. Organic farmers work hard to develop healthy soil—and then work hard keeping it healthy.

Organic farmers use organic compost from recycled plant and waste materials. When cow or chicken manure and food waste/scraps are composted, microorganisms (bacteria, earthworms) break down and digest raw components of the waste. This metabolic activity generates temperatures of 131F to 149F degrees, which in turn kill any disease-causing bacteria and weed seeds.

## Natural Minerals

Preparing a good seedbed to grow tobacco is key.

Organic farmers sometimes add natural minerals that help improve soil consistency and pH balance. If the farmer needs to lower the soil pH (make it more acidic), natural, mined elemental sulfur may be added. To raise the soil pH (make it more basic), powdered limestone may be added. The "ultimate pH range" for most vegetables is between 6.0 and 7.0.

Citric acid can be added to water used in an organic tobacco floatbed.

During transplanting, growers often add fish emulsion to the transplant water. Make sure to peg-in plants to achieve maximized plant population.

Growers rely on soil tests to prescribe the exact nutrients required in the soil to grow a quality crop of organic tobacco. The basic principles of fertilization are the five R's:

Apply the Right nutrient
At the Right rate
At the Right time
In the Right place
And at the Right cost

Kentucky burley grower Chris Korrow says the key to growing a successful organic crop is keeping the nutrients in his mountainous soil at the proper levels. That is achieved in part by increasing organic matter to high levels in the soil. So he uses compost. He also uses two cover crops—buckwheat and sweet clover. Rotating crops is particularly important in growing organic tobacco.

We'll have more about rotation in a following section, but let it be noted that, for Chris, that means growing organic garlic in alternating years with tobacco. Bacteria and fungi are also important to hold nutrients in the soil, encouraging better soil drainage, giving better soil aeration and helping the tobacco crop resist drought. Nitrogen for organic burley may be found in several sources—one being composted manure.

To grow a successful organic tobacco crop without using prohibited chemical fertilizers and pesticides, growers follow SFNTC's successful program of alternative fertilization techniques and pest control methods. Developing and implementing a conscientious soil-building program enhances organic matter and encourages optimum soil health. Soil must be tested and lime applied to gain the proper pH balance in the soil, as noted earlier. There are good natural sources of potash, phosphorus and lime. Composted poultry manure, fish and blood meal are ideal sources of nitrogen. Blood meal or fishmeal makes a good side dressing. Minerals used are natural phosphates, potash, bone meal or seaweed extracts.

Field test in tobacco field by North Carolina Department of Agriculture and Consumer Services.

To increase nitrogen after transplant, South Carolina grower Ben Williamson used a teaspoon of blood meal on each individual plant. "This practice also kept deer away from the tobacco crop," he said. Ben said deer were his worst problem. "Sometimes the deer bit the tobacco plants off, and other times they pulled the plants up and left them on top of the ground. One year we saturated cotton bolls with coyote urine and scattered them throughout the field. Combined with the use of blood meal, this was a very effective practice."

The Ball brothers used organic fertilizers, such as fish and bone meal. That practice came about as a result of working with the North Carolina Department of Agriculture in a three-year field test that addressed fertilizer needs for organic tobacco. The Ball brothers and many other growers had assumed organic fertilizers needed to be applied several weeks before planting so the material would have time to break down and release nitrogen. But the research showed that if growers put out the fertilizer in March, then much of the nitrogen would already be gone by the time plants were set out in May.


## The Best Time to Apply Organic Fertilizer

Applying organic fertilizer about two weeks before planting has worked well. Guidelines from the study noted above have helped growers decide how much fertilizer to apply before planting and how much to add later as side dressing.

When Alamance County, North Carolina, farmer Jane Iseley started growing organic tobacco, there were no established guidelines available for fertilization. She took advantage of plant tissue analysis and advice available from her regional North

Carolina Department of Agriculture agronomist. In growing organic tobacco, tissue testing has two important uses—it enables the grower not only to fertilize more precisely, but to time-harvest more precisely as well. Jane and other growers found that organic fertilizers cost three times as much as conventional fertilizers. "By using tissue analysis to adjust fertilization rates, I was able to produce quality leaf as well," Jane said. "Without it, I would have harvested a week earlier."

Because of limited fertilization, Granville County, North Carolina, grower Lane Mize admits that his organic leaf doesn't look as leafy or robust as his conventional tobacco that was planted in a nearby field the same day. He attributes the difference to the slower release of chicken manure into the soil around the organic plants versus the ammonium nitrate that is used with conventional tobacco. "My organic tobacco is two weeks behind," Lane says, "but that doesn't worry me, because I can't harvest all my acres at the same time anyway."

Orange Country, North Carolina, grower Stanley Hughes says the fertilizers used in his organic fields are all-natural mixtures of dried bone meal and chicken litter, as well as seabird guano.

Ray Watkins, a farmer based near Oxford in Granville County, North Carolina, first limed his fields to help build up nutrients in the soil. He also applied bagged chicken litter as his main source of nitrogen. To get some extra nitrogen after transplant, he puts a teaspoon of blood meal on each plant. He uses Sul-Po-Mag and bone meal as sources of potash and potassium. Ray said there's a drastic difference in how he'll fertilize his organic crop compared to what he puts on his conventionally grown tobacco.

He has also planted sunflowers as a buffer between his organic and chemically treated crops.

Ray Watkins, seen here in an earlier photo with his family, has used lime, bagged chicken litter and other approved materials to build up nutrients in his soil.

## Controlling Weeds

Controlling weeds is a major challenge for both organic flue-cured and burley tobacco growers.

Successful weed control must be accomplished with cultivation, as no prohibited herbicides are acceptable for organic production. Several cultivations and hand-hoeing will help keep weeds in check through the critical growing period. Shallow cultivation helps provide aeration and rebuilds a high, wide bed.

Ray Watkins' father cultivates for weed control.

Weeds were controlled by cultivation on Ben Williamson's farm. He used a rolling cultivator after the first rain, usually one week to 10 days after setting the plants out. Workers then hoed thoroughly around the plants once. Less intensive weeding took place as needed. Weeds fell to a cultivator and with regular trips to the field with hoes. Maintaining a good plant row canopy shades out weeds and is another good method of weed control.

If Ben had a problem with weeds in one of his fallow fields, he planted a summer cover crop of millet or another "smother crop." "Nut grass and Bermuda grass were effectively controlled this way," he reports.

Successful farmers use cover crops as much as possible. Such crops as sunflowers and marigolds attract beneficial predatory insects that feed on pests. Typical cover crops that are grown and then tilled into the soil include Austrian field peas, bell beans and vetch. These cover crops add nutrients, such as nitrogen as well as organic matter, to the soil.

"Since we can't use prohibited herbicides," says Billy Carter, "we have to do a lot of cultivation from the start. It takes some time, but on a small acreage it's not too much trouble."

Organic tobacco requires more manual labor than conventional growing, as Tommy Winston, center, and two of his workers can attest.

Stanley Hughes says, "In conventional tobacco, you can take your tractor and spray and spray. I could spray probably 10 times where we're now doing maybe one or two times in our organic fields with approved materials. With weeds in conventional tobacco, you can use a herbicide to help keep out the choppin' hoe, but with organic tobacco, you're going to have to have more labor to do that."

Burley grower Chris Karrow manages weeds and grass using a summer buckwheat cover crop and manure. He finds this also makes the land easy to cultivate and hoe.

For Ray Watkins, growing free of prohibited materials means he uses several cultivations and hand hoeing by his H-2A (immigrant) laborers to control weeds in his organic fields. "I'm planning on putting a lot of labor into my organic crop," he says.

*Special reminder: Using plant row canopies to shade out weeds is also an effective method.*

## Controlling Insects

Along with weeds, insects are the biggest pest to organic growers.

Protecting the leaf from the greenhouse to the warehouse is vital for an optimum, quality crop. Planting companion crops that attract and favor many beneficial insects and birds helps control insects, along with organically approved products. Planting sunflowers alongside the tobacco crop helps with aphid conrol, as do organic insecticidal soaps. Budworms are controlled by placing a pinch of Bt bait in each bud.

Organic growers are pleased with the pest control they receive from beneficial insects like predator wasps, lacewings and ladybugs. Birds are a friend of the grower, too; birds carry off hornworms up to two inches long. Farmers also learn that sterile fields devoid of pests or predators can prove to be an expensive liability. Maintaining a good natural balance between pest and predators saves significant money.

Aphid infestation is more common in conventionally grown tobacco, says Ben Williamson.

Ben Williamson planted sunflowers alongside his tobacco to control aphids. Rather than plant sunflowers just around the field, Ben planted two rows of tobacco, held one turn row for equipment, planted another four rows of tobacco, held another row for equipment, planted two more rows of tobacco and then planted four rows of sunflowers. He repeated this pattern until the required number of tobacco acres was planted.

When the tobacco got knee-high, the sunflowers, planted in early March, were the same height as the tobacco. The sunflowers attracted and nurtured lady bugs, which moved into the tobacco and ate the aphids. "Aphids don't seem to bother organic very much," Ben said. "During one span of three crop years, I had a tremendous infestation of aphids. While they damaged my conventional tobacco a great deal, they didn't stay very long in the organic tobacco and did very little damage to the crop."

Ben found that hornworms succumbed to wasps. So he used two species in this tobacco. "Red wasps are great friends of tobacco. They lay eggs on the hornworms, spin cocoons on them and feed off the worms. The worm quits eating and dies when the wasp larva is on its back. Yellow and black wasps (the ones that sting) are also attracted to organic tobacco. I've seen the wasps carry off the hornworms up to two inches long."

When natural predators don't get the job done, growers can use a bacillus thuringiensis (Bt) bait, a naturally occurring pesticide, to control worms. Birds also help control the hornworm population. By August, the sunflowers produce seeds and attract the birds. But the hornworms are an added treat. Ben found mockingbirds, brown thrashers and blue grosbeaks very helpful in organic tobacco production.

Populations of nematodes (tiny, underground pests) are reduced by growing white vetch, which doesn't host them. Ben, however, sometimes used a winter cover crop of Austrian winter peas to help control weeds, which do host nematodes. "I did not have a difficult time with any of the diseases typically found in a conventional tobacco field," Ben said. Blue mold, for example, did not affect Ben's organic tobacco crops. But he occasionally had a problem with tobacco mosaic disease. To deal with that, he quarantined the plants and did not sucker them. He also saw only a one-percent loss from Pythium root rot and black shank.

Billy Carter, who grows 42 acres of organic and 200 acres of PRC tobacco, knows that aphids or diseases such as blue mold or black shank can quickly decimate an entire crop.

Ladybugs are attracted to tobacco plants when sunflowers are planted nearby—and the ladybugs make a meal of aphids.

"With organic tobacco, minor things can get totally out of hand and you can lose a whole crop 100 times faster than you can a conventional crop," Billy says. "When you have a problem, it can be magnified greatly because you don't have a full arsenal to go after it."

Instead of using prohibited pesticides, Billy also plants sunflowers in his organic tobacco fields to draw ladybugs. These eat aphids, which "like to suck the life out of tobacco," he says. To combat caterpillars, Billy uses the naturally occurring pesticide Bt.

The Ball brothers had to do some hand worming, 'glugging' on soybean oil meal to the crotch of the leaves and stalk. But pest control was no serious problem in their first growing season. Like many others, the brothers planted sunflowers around their tobacco fields to control aphids, which they found particularly troublesome in the early season.

Virginia organic farmer Ralph Tuck says growers must plant sunflowers earlier because they take 60 to 75 days to flower. Buckwheat can be planted later because it flowers in 30 days. For worm control, Ralph also uses Dipel, a Bt product, which is approved for use on organic tobacco.

About aphids, Ralph says they do damage at the bottom of the leaves. "Aphid droppings are what destroys tobacco and fall on the lower leaves."

## Irrigating Organic Tobacco

During dry weather, organic growers can irrigate their

tobacco, provided, of course, the water does not contain any prohibited chemical residues.

Irrigating the fields.

## Topping and Suckering Organic Tobacco Plants

Because the use of chemical suckering agents is not allowed with organic tobacco, topping and suckering are the most time consuming but important tasks associated with growing organic tobacco.

The aim of sucker control is to focus the plants' energy into filling the leaves rather than growing the flower. Because tobacco sells by weight, heavier leaves mean a bigger paycheck.

Suckers—branches that a tobacco plant likes to sprout and which take energy away from the leaves that the farmer wants to harvest—normally are treated in conventional farming with a chemical that burns the sucker off when it starts to emerge. It is sprayed from a tractor. In an organic field, the farmer drips cooking oil on the sucker branches. It has the same effect as the chemical, but must be done by hand, plant by plant, because the oil won't go through a sprayer.

Billy Carter tops (the top flower) of his organic tobacco plants by hand to promote growth.

Early topping to improve yield and quality is usually done by hand. Suckers can be removed by hand as well as stunted by carefully applying approved soybean or mineral oil to the top of the plant. The farmer must be sure the oil runs down the stalk and into each leaf axil to get good control.

"Topping and suckering are the most time-consuming tasks associated with growing organic tobacco," Ben Williamson said. "We did it every week for 10 weeks, and it took one person per acre to do the job." Normally, Ben never allowed his plants to flower. Instead, workers topped the plants at 15 leaves. He then suckered the entire plant, leaving one sucker in the top, which produced eight to 10 leaves before it was topped.

While the Ball brothers' concern about dealing with insects without insecticides and weeds without herbicides did not prove to be a big problem, they did find controlling suckers a challenge. "We jugged on (applied by hand) mineral oil, and that helped to knock the suckers back, but it had to be done mostly by hand," Randy said. The Balls and other growers found that controlling suckers when they are small is advantageous. "If you can keep them cleaned out when they are small, sucker control is much easier," said Randy. The Balls did their initial topping and suckering by hand. They then poured vegetable oil from gallon jugs over each plant, allowing the oil to run into each leaf axle, all the way to the bottom of the plant. The vegetable oil worked like a contact sucker control agent, Randy said. "We used either vegetable oil or mineral oil. Both of them did a good job of controlling suckers."

A young Eric Ball suckers the plant, also by hand.

Ralph Tuck tops his crop early, because once the tobacco is topped, he says, aphids usually don't bother the tobacco much. He also advises to "irrigate and get it topped as quickly as possible, because a small window of opportunity exists between lay by and topping." Besides topping early, Ralph also does not plant too close to woods to keep the insect population away from his tobacco. "It's like a draw to them," he says. He also notes that smaller fields seem to attract more aphids. In larger, open fields, airflow is better and he has seen fewer insects in those fields.

Stanley Hughes applies a blend of soybean oil and mineral oil to the heads of the plants to prevent flowering. Stanley also estimates that 18 hours can be spent per acre removing suckers.

Billy Carter relates another interesting experience, telling how he goes to a local discount store to buy gallon jugs of soybean oil. Turns out it is the same soybean oil that is used at home for cooking in the kitchen.

## Coping With Plant Diseases

Let's not mince words. Plant diseases are tough to control for the organic tobacco farmer.

Since few approved disease-control materials are allowed in organic tobacco production, the challenge is to effectively use varietal and resistant plants, proper crop rotation and good sanitary practices to keep diseases from appearing in the first place.

There are no known organic control agents for blue-mold, black shank and Granville wilt.

Grower Billy Carter says a conventional tobacco farmer can deal with many common tobacco diseases with relative ease, but there is not much an organic farmer can do about them. "If you get those diseases, such as black shank or Granville wilt, in organic tobacco production, you have no options. You lose the crop, essentially," he said, which makes careful attention to organic farming practices all the more important.

Kentucky organic burley grower Chris Karrow says healthy soil helps control black shank and other diseases. Rotation fits in well with the soil-building concept of organic, so successful organic farmers tend to be resourceful in this area. And the rotation must take as long as it needs to. "We had one field that had some black shank, so we left it out of tobacco for five years," Chris says. "That seems to have taken care of it."

Burley grower Roger Smith offers that TN-90 and TN-997, types of plant seeds, have been popular among organic growers because of their disease resistance.

The answer to this potentially difficult situation is pretty straightforward for Virginia grower Ralph Tuck: "Controlling diseases means relying on resistant varieties and rotation."

## Harvesting Organic Tobacco

Organic tobacco is often harvested by hand. As workers harvest the four bottom leaves, others continue late topping and suckering of plants. Most growers will harvest three or more times, based on curing barn capacity and the weather. At harvest time, yellowing agents used to cure tobacco are not allowed in the field or the curing barn.

South Carolinian Ben Williamson harvested his entire crop by hand, saying, "We took three to five leaves at a time and made five or six passes through the field."

Like all of his fellow organic growers, North Carolinian Ray Watkins takes great care in having his workers keep the organic leaves separate. Ray's harvested leaves are cured in separate tobacco barns from his conventionally grown tobacco.

Organic tobacco was hand harvested and placed in separate curing racks on Ben Williamson's farm.

## Curing Organic Tobacco

While quality tobacco is made in the field, it can be either retained or lost during curing. That means controlling relative humidity and temperature are absolutely essential in the curing process. This rule is as true for organic tobacco as it is for conventional.

What is different is that organic tobacco must be stored separately from conventionally grown tobacco, and signage must note that it is an organic area. So organic growers must make sure a responsible person oversees tobacco going into and out of the curing barn. Keeping good records is essential!

There are different approaches to curing flue-cured and burley tobacco. For flue-cured tobacco, the leaves are placed into racks for curing. In the curing barns, heated air is forced through the tobacco to cause it to first turn yellow or orange in color, and then to dry the leaves and stems. The initial temperature for yellowing the leaves is 95F to 100F degrees, and is then increased incrementally to 165F to 170F degrees. Ventilation is part of the curing process and is varied as needed to remove moisture while retaining quality of the tobacco. It takes five to seven days to cure a barn of flue-cured tobacco.

For Ben Williamson, curing was accomplished for years using a wood-fired boiler to heat water, which was then circulated through the rack barns. Only in his last couple of years of farming did Ben turn to liquid gas.

Burley tobacco is cured much differently than most tobacco grown in the Carolinas, Virginia and adjacent states.

Burley is cured by air. It is hung in unheated barns and allowed to cure over a period of weeks rather than days. For Kentucky grower Chris Korrow, air-curing takes 45 days, after which Chris' friends help him strip and bale the crop.

In addition to flue-cured tobacco, Billy Carter grows burley tobacco on his North Carolina farm and then air-cures it the traditional way.

## Preparing Organic Tobacco for the Receiving Station

SFNTC's contract organic tobacco growers deliver their cured tobacco bales to receiving stations at the company's manufacturing plant in Oxford, North Carolina. There, a certified tobacco grader evaluates each tobacco bale and applies an SFNTC-designated grade. Each bale is probed for a sample to be tested, for both moisture content—it must not exceed a certain amount—and residues.

At the receiving stations, tobacco is certified by graders, evaluated and probed for a sample to be tested for moisture content and residues. Willy Brooks, left, Kirk Gravitt, center and Randal Ball, SFNTC's leaf manager.

When it comes to grading, SFNTC uses a USDA standard. This includes stalk position, quality and color of the tobacco—the three key elements.

After weighing, each tobacco bale is probed for a sample that is sent to an outside, independent laboratory to confirm that there are no prohibitied residues in the tobacco.

The sample taken from each bale is analyzed against a three-page list of not-allowed chemicals. We have, as required by the USDA organic program, a near-zero tolerance policy.


## Post-Harvest Cultivation Practices

All the SFNTC burley growers are encouraged to destroy stalks and roots just as soon after harvest as they can. Left in place, these stalks and roots are a perfect place for breeding diseases. This is also the time to start planning the next rotation crop to keep fields in organic production.


## Rotation With Other Organic Crops

As noted throughout this chapter and in other parts of this book, maintaining healthy soil and an effective crop rotation program are critically important for organic tobacco growers.

Planting tobacco over and over in the same field depletes the specific nutrients needed by the tobacco plant. So organic farmers carefully rotate their crops every season to avoid this condition, known sometimes as "soil lock."

Our organic growers have learned some good lessons on how to make their soil more productive. Those include performing regular soil tests and adding, as needed, various nutrients that can impact the crop and the soil itself.

Basically, if a certain crop depletes nitrogen, organic farmers will introduce crops, like peas and beans, for example, that add some nitrogen to the soil the next season. On Ben Williamson's farm, organic tobacco was grown every fourth year on his certified land. "In the intervening years, I planted a cover crop one year and organic soybeans were planted the other two years."

Organic burley tobacco grower Roger Smith also cultivates specialty corn, poppy seeds, zinnias and marigolds as his tobacco rotation crops in Kentucky.

In addition to growing certified organic tobacco, Alamance County grower Jane Iseley raises tomatoes, corn, strawberries, beets, cabbage, broccoli, snap beans, pumpkins, peppers and squash (both winter and summer varieties). She currently has 90 acres of organic crops, 20 of which are organic tobacco for SFNTC.

Stanley Hughes grows several other organic crops, including sweet potatoes, collards and corn. "Over the past several years, I also began growing more organic vegetables as well, including cabbage, collards, kale, beans and squash for 'tailgate marketing' at the nearby Durham and Carrboro farmers' markets."

Richard Ward has a total of 135 acres of certified organic land. In addition to tobacco, this Whiteville, North Carolina, farmer cultivates sweet potatoes, melons, collards, broccoli, strawberries, squash and other fruits and vegetables.

Many organic tobacco growers rotate crops to promote nutrient-rich soil. Jane Iseley grows organic produce for sale at her produce stand.

Along with his organically grown tobacco, Granville County grower Tommy Winston raises five acres of trellised tomatoes. In addition to replenishing his North Carolina soil, this provides work for his H2A laborers during the period just before tobacco harvesting begins.

As a further soil-building technique and for disease prevention, Kentucky organic burley grower Tom Croghan follows a four-year rotation, typically growing tobacco two years, then alfalfa and orchard grass for two years. As a result of using minimum tillage techniques, Tom also finds that he now has very little soil erosion in his fields.

Richard Ward grows and markets a variety of organic vegetables.

# Other Earth-Friendly Tobacco Growing Practices

As a processor and manufacturer of 100 percent additive-free natural tobacco for more than 25 years, our commitment to earth-friendly products and the land from which they come runs deep.

Quite often, that has meant doing things differently.

SFNTC began exploring alternative ways of growing tobacco at the outset. Initial efforts in the early 1980s to grow tobacco in the Southwest, on American Indian reservations in New Mexico, did not work out—the land and climate were not right. So the company sought to purchase only the finest, natural tobacco available where it does grow.

This led to working with growers, agricultural agents and scientists at state universities in the Southeast.

The result was our organic program, which is outlined in these pages and chapters, and our line of certified organic tobacco products. That effort and commitment is a reflection of our earth-friendly values.

And it led to the development of an overall growing approach that is reducing the use of pesticides on the farm and promoting other sustainable practices.

Spraying of certain approved materials is allowed for PRC tobacco.

In 1991, we began what is now called our purity residue clean (PRC) program—a program to reduce pesticide residues on our non-organic, additive-free line of products. We pay the farmers a premium to not use certain systemic chemicals that leave residues on the leaf and promote sound cultural practices prescribed by the company. Most growers find the practice a better way to produce tobacco. Through the success of this program, we are using more PRC tobacco every year.

Ronald Stainback grows PRC tobacco, a program that SFNTC began in 1991.

Growers under our PRC tobacco program have eliminated the use of such chemicals as Temik, MH30, Endosulphin and Prime Plus in their fields.

## Chemicals can Last a Long Time in Soil

Richard Ward learned early on how long-lasting chemicals can be. Before going organic, he grew PRC tobacco for SFNTC.

One fall early on, he got a call from SFNTC's leaf director Fielding Daniel who told him the company had found traces of a chemical not allowed in the PRC program in his leaf that had been brought in.

Richard Ward was surprised to find out how long chemicals can stay in the soil.

"I never gave a thought that I'd have a chemical problem since I hadn't used the land in years," Richard says. So he checked and re-checked his records. "We went back through four years of crop reports, which showed we had not done anything wrong, and certainly didn't use that chemical.

It turns out that the field that had produced the tobacco leaf in question had a sandy soil over clay soil, which can hold a lot in the ground. Seven years earlier, the chemical had been used. Seven years. "The chemical company, when I called them, said the chemical breaks down and is supposed to be gone after four years," he says shaking his head.

The Stainback farm.

## PRC Tobacco Certification Requirements

Growers of PRC tobacco are able to use soil fumigants, commercial fertilizers, and other approved pesticides in a limited manner, to assure that no chemical residues remain on the cured leaf when brought to market. Producers must leave a buffer strip from any field in which chemicals are being used—a 100 foot clearance from crop rows of any conventionally grown tobacco, and 500 feet from any area where non-PRC chemicals have been mixed or stored within the last three years. There should be no indication of surface water run-off from non-PRC fields.

In the PRC program, growers must detail all cultural methods and practices on the PRC field over the last three years. Initial soil sampling is used to provide a comparison, should any low-level background pesticide residues be detected later in crop samples. An initial site inspection and grower orientation is conducted under the direction of an environmental analyst hired by the company.

At least two unannounced inspections by an SFNTC environmental analyst will occur during the growing season to take green tissue samples from the field for analysis. A final product sample is collected at the receiving station for pesticide residue tests, leading to PRC certification of the tobacco crop.

"Growing PRC tobacco is like we used to grow," says Sam Crews.

Safe-use chemicals, which break down quickly, can only be applied under the program when levels of pests have reached thresholds recommended by the county extension agent in the grower's area. Spray tanks and equipment must be washed of all previous chemical residues using a detergent. Tank mixes must include the lowest effective concentration of safe-use chemicals.

"Growing PRC tobacco is like we used to grow," says Oxford, North Carolina, tobacco farmer Sam Crews, who grows some 30 acres of PRC tobacco, in addition to 120 acres of conventional tobacco.

### Requirements and Process for Growing PRC Tobacco

While we work closely with our PRC tobacco growers, both our PRC and conventionally grown tobaccos are bought for us by our leaf dealers—representatives from both United Tobacco Company and Universal Leaf Tobacco Company. Our leaf dealers contract on our behalf with tobacco farmers to grow a certain amount of tobacco each year exclusively for us.

This arrangement allows us to maintain a close relationship with the growers.

### Growing Seedlings for PRC Tobacco

We suggest our growers use conventional float-house grown transplants to develop uniformity in the crop. We ask they be sure to select the seed variety that best fits the needs of the grower's specific growing or marketing conditions. At transplanting, growers may use water-treatment chemicals that are approved for use to control early season insects.

PRC grower James Bing with seedlings.

## Land Preparation and Fertilization for PRC Tobacco

Land preparation is largely the same process as with organic tobacco. To fertilize a PRC crop, any formulation normally used on tobacco that does not contain a herbicide is acceptable. Our environmental analysts recommend growers closely follow soil test recommendations using manure-source nitrogen when possible and phosphate only as indicated by soil tests. They should apply a sufficient amount of nutrients at planting to provide for normal growth of a successful crop. Nitrate of soda is acceptable as a side-dressing.

A few more words about fertilizers:

The three major nutrients that all plants require are nitrogen, phosphorus and potassium, which are the basic nutrients found in standard agricultural fertilizers. Specific fertilizer formulations will vary from grower to grower, depending on soil tests conducted by an agricultural extension agent. Levels of nitrogen, phosphorus and potassium indicated by the soil analysis will generally dictate the recommended fertilizer formulation. Lime also may be applied, as recommended.

Tobacco crops deplete soil nutrients at a much higher rate than most other crops; as a result, conventional tobacco farmers use an abundance of fertilizers. The crop rotation practices used in traditional farming minimize the depletion of soil nutrients.

In the mid-1900s, the tobacco industry determined that fertilizers high in phosphate would produce even higher-yielding tobacco crops. In fact, about 85 percent of the soils on which tobacco is grown have very high levels of phosphorus. This nutrient has built up in tobacco fields due to continued use of high phosphate fertilizers. Apatite, which is a group of phosphate minerals, is used by many to produce "super-phosphate" fertilizers. SFNTC's contract growers do not use "high phosphate fertilizers" in growing any of our tobacco. In fact, we've found that additional phosphorus is *not* essential for the growth of tobacco.

On the other hand, the basic N-nitrogen, P-phosphorus, and K-potassium-based fertilizers (standard and custom-blended) may be used in the production of both our PRC tobacco and conventionally grown tobacco, but they are not "high phosphate fertilizers." These fertilizers may also contain other essential nutrients.

While Jimmy Crews uses a variety of materials on his conventional tobacco, he is limited under PRC.

### Weed Control in PRC Tobacco

Reduced use of costly herbicides to control weeds dictates an increased cultivation schedule and perhaps a once-over hand-hoeing. The SFNTC PRC program has approved certain chemical treatments to help control early season weeds and grasses. After tobacco is laid-by, a final cultivation is in order.

### Insect Control in PRC Tobacco

Our PRC program also includes approved chemicals that provide excellent control of flea beetles and early season aphids. If nematodes are a problem, the program allows the use of an approved nematicide with the provision that it only be used at least two months prior to harvest.

### Irrigation of PRC Tobacco

Irrigation of a PRC tobacco crop may be essential for crop maturity during dry weather, provided of course, the water does not contain any prohibited chemical residues.

### Topping and Suckering PRC Tobacco

We recommend tobacco topping at a maximum of 18 leaves to gain full leaf maturation. This also assures fewer sites for sucker development. MH-30 or Prime-Plus cannot be used for topping and suckering PRC tobacco. Instead, topping should be followed immediately with an approved contact suckercide containing only fatty acids as the active ingredient.

The flowering top of a tobacco plant is "topped."

One week to 10 days following topping, the plants should be carefully hand-suckered and the contact suckercide re-applied. Growers should count on re-applying the suckercide about three times after topping.

**Harvesting PRC Tobacco**

Proper ripening of the tobacco demands at least three harvests. While most small farmers usually harvest by hand, larger growers often use mechanical harvesters to bring in their crop. No ripening agents in the field or barns are acceptable for use on PRC tobacco.

PRC tobacco must be cured separately from conventional tobacco.

## Curing PRC Tobacco

PRC tobacco must be stored separately from conventionally grown tobacco. A responsible person must oversee the tobacco going into and out of the curing barn. Cured tobacco should be ordered with only sufficient moisture to prevent shattering during handling and baling.

As noted with organic tobacco, controlling relative humidity and temperature are absolutely essential in the curing process.

For flue-cured tobacco, the leaves are placed into racks for curing. Heated air is forced through the tobacco in the curing barns to cause it to first turn yellow or orange in color, and then to dry the leaves and stems. The initial temperature for yellowing the leaves is 95F to 100F degrees, and is then increased incrementally to 165F to 170F degrees. Ventilation is part of the curing process and is varied as needed to remove moisture while retaining quality of the tobacco. It takes five to seven days to cure a barn of flue cured tobacco.

Burley is air cured in unheated barns.

With burley tobacco, curing is accomplished by air as the tobacco is hung in unheated barns and allowed to cure over a period of weeks rather than days.

## Preparing PRC Tobacco for Delivery

Delivery of PRC tobacco to the SFNTC receiving station should be in large, 400-pound tobacco bales. A certified grader evaluates each bale, applying a SFNTC designated grade. After weighing, each tobacco bale is probed for a sample, which is sent to an outside, independent laboratory to confirm that there are no measureable chemical residues in the tobacco.

PRC tobacco is brought to receiving stations in 400-pound bales, where it is evaluated, here by Randal Ball.

On market day, growers receive a check for the market value of their tobacco. After the tobacco has been certified "Purity Residue Clean," SFNTC pays the agreed upon premium to the current market value of the crop.

PRC growers Lynn Vick, left, and David Rose, right, with SFNTC's Fielding Daniel.

## Post-Harvest Cultural Practices for PRC Tobacco

At the end of the season, stalk and root destruction and the planting of a cover crop are highly recommended. That is also the time to start planning crop rotations, an effective method of disease control for successful production of tobacco.

## Rotation With Other Crops

As noted in previous chapters, maintaining healthy soil and an effective crop rotation program are critically important for tobacco growers. Planting tobacco over and over in the same field depletes the specific nutrients needed by the tobacco plant.

PRC growers have learned some good lessons on how to make their soil more productive—performing regular soil tests and adding, as needed, various nutrients that can impact the crop and the soil itself.

If a crop depletes nitrogen, farmers can introduce crops, like peas and beans, for example, that add some nitrogen to the soil for the next season.

# PRC Growers Talk About the Process

Over the years, a number of our Purity Residue Clean (PRC) growers have talked about this earth-friendly process. We share some of those thoughts below.

**Sam & Jimmy Crews**, Oxford, Granville County, North Carolina

*In an effort to diversify their tobacco growing operation, the Crews brothers contract some of their crop with nearby SFNTC to grow PRC tobacco. Sam offers his perspective:*

"We've been growing PRC tobacco for SFNTC for about six years. We started out with six acres and this year we have about 70 acres contracted. It's been a good experience. It's a little different than the conventional—you can't use quite as many chemicals. It takes a little more management to do it.

"We're basically doing what I did growing up on my daddy's farm. We plant the tobacco and have very few chemicals we can use. Sucker control is the biggest challenge. We have to go over it twice for sucker control. We have to store it separately and have to make some good decisions about harvesting so we can fill one barn at a time. The premium SFNTC pays us is based on the quality of our tobacco."

Brothers Jimmy Crews and Sam Crews

**Ronnie Perry**, Rolesville, Wake County, North Carolina

"We tend about 160 acres and I've got about 40,000 pounds contracted with Santa Fe. I think I've been raising PRC tobacco for SFNTC for five years. Some chemicals you can't use and you have to find a different way for your sucker control. I like to do it all like that. I'd love to grow more PRC tobacco and grow some organic, too."

**Ronald Stainback**, Warren County, North Carolina

"We've been growing PRC tobacco for SFNTC for several years. It's been a real good working relationship. The premium on PRC tobacco really helps us out these days. We grow about 100,000 pounds of PRC tobacco for SFNTC. My sons and I operate a receiving station for SFNTC through Universal. We receive the PRC tobacco in this area here. It's been a real good working relationship, too."

**Thomas Dean**, Wake County, North Carolina

"The yield was outstanding as far as I was concerned; we made 103 percent on it. The quality was right there with the rest of it. We were really satisfied with it. On this 150 acre operation, we've got enough workers to grow 10 to 15 acres of PRC tobacco. The SFNTC guys that come out and check the fields have always done what they said they would do and I tried to do what I said I would do. It works out real well."

**David and Allen Rose**, Nash County, North Carolina

"We're growing more than 100 acres of PRC tobacco, up from 30 acres when we started contracting with SFNTC via United Tobacco. We feel we're getting a better deal with Santa Fe, and even better, we've enjoyed working with them. We grow a total of 500 acres.

"With PRC tobacco, we have to keep everything separate

from our conventional crop, as we can't have any chemical residues showing up in the crop we take to market. We just have to be extra careful to make sure everything is cleaned out well when we go from one type of tobacco to another. We can't afford to contaminate our PRC tobacco. There's too much of a premium to lose for it to become a conventional crop.

"One of the things that attracted us to PRC was that we use hand labor. We do a hand sucker about two times more than using MH30. We feel like the yields are almost equal. So we don't see a lot of difference between PRC and conventional tobacco. We've been pleased with SFNTC's grading schedule, we feel like its been graded fairly, and we feel like with the premium we're receiving now, it's worth the extra effort to raise a crop without some of these other chemicals. We feel we will grow more PRC in the future because we feel this is a way to utilize our labor better."

Brothers David Rose and Allen Rose.

**Lynwood Vick**, Wilson County, North Carolina

"We're a family operation. We raise 380 acres of flue-cured tobacco, and 120 acres this year was PRC. We've been raising PRC tobacco for SFNTC for five years now. When we first started growing PRC, we were kind of leery of how the crop would be grown without the use of some of the major chemicals we use in conventional flue-cured production.

"We started off with 10 acres the first year and have doubled it every year since then—because we feel it's just as easy to grow and actually the quality of the tobacco is much better than where we do use a lot of chemicals on it. We do have to use more labor on it, but it seems like the amount of yield and the price we get for our tobacco from SFNTC compensates for our extra labor cost that's involved. That's been a big positive in growing PRC tobacco."

Lynn Vick

**Tim Shelton**, Wilson County, North Carolina

"One thing about SFTNC. They're a company that tells straight up what they can and will do, and they hold up their end of the bargain. It's been a pretty good relationship with SFTNC. It remains one of the best games out there for a small farmer like me."

**Donny Blizzard**, Snow Hill, North Carolina

*Donny has grown the PRC crop four years and in 2007, he grew 217 acres (more than 70 percent of his total crop).*

"Growing PRC tobacco is a very interesting concept. We've learned quite a bit from it, especially that there are some cheaper chemicals that we can use that have a lower impact on the environment and still get 95 percent control of insects compared to using more expensive products.

"I wouldn't have learned that without growing PRC. Last year, when the thrips population wasn't real bad, we got the same disease control using both products. Where we had water to irrigate the crop, I had some of the best tobacco I've ever grown—in terms of high quality and yields. If you are growing a lot of tobacco, you need to be involved in making it a big success."

Donny Blizzard

# From Farm to Finished Product— Organic Tobacco and the Manufacturing Process

From the time organic tobacco leaf leaves the growers' hands until it rolls out of the manufacturing plant as finished product, a complete, thorough and fully documented audit trail ensures the tobacco's organic integrity.

Extreme care and attention to detail follows the organic tobacco leaf. To help ensure that safe passage, we have put together a comprehensive planning and operations guide, our *Organic System Plan.*

The plan and its many respective components are a guide not only for those who process and manufacture tobacco, but also for the many suppliers, growers and a host of others who are responsible for one or more aspects of helping render a finished product. The plan book also serves as a repository for the numerous certification documents and applications, including official USDA certification documentation and certification requirements. It is made available to the USDA and the National Organic Program and their inspectors.

More on the Organic System Plan later in this chapter, but, first, let's follow the progress of the organic tobacco leaf.

**Processing & Production**

Once harvested and cured, organic tobacco is ready to be delivered to a receiving station, where it is measured, weighed and tested. It is then transported to a stemmery for processing.

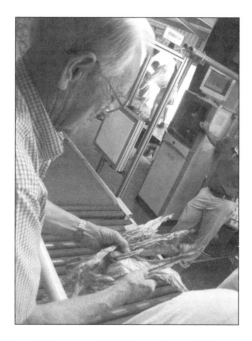

Once at receiving stations, organic tobacco is graded, measured, weighed and tested, in this case, by Willy Brooks.

For SFNTC, this means using our two contract leaf dealers who de-stem the tobacco leaves. Unlike almost all U.S. manufacturers, SFNTC uses only whole leaf, or the lamina, of the tobacco plant. We do not use stems or scraps that bulk up and result in less tobacco leaf in the finished product. After we remove the stems, tobacco strips are compressed into bales and stored in 400-pound boxes, which we transport to our Oxford warehouses for two or more years of aging, a process where the moisture content is standardized.

After stems are removed, tobacco strips are compressed, stored in boxes and sent to our warehouses in Oxford for two or more years of aging.

After aging, the organic tobacco is ready to be blended with other organic tobacco and made into finished product.

For us that is either loose tobacco for cans and pouches, or for finished cigarettes. We produce our canned and pouch tobacco at our West Street facility in Oxford. Cigarettes are produced at our Knotts Grove plant, also in Oxford. Both facilities are certified organic handlers.

After aging, organic tobacco is prepared for blending.

## Production Begins—in *Primary*

When we are ready to start blending a new batch of tobacco, boxes of various grades, years and types of tobacco are brought into the *primary* processing area at Knotts Grove.

The tobacco is brought to our *primary* processing area and conditioned for blending.

To produce a blend that delivers our desired flavor profile, we use various types and styles of organic tobacco. After the baled tobacco is sliced into smaller chunks, it enters a conditioning chamber where we add steam and filtered water to increase the tobacco's moisture content. From there, precise amounts of the various tobacco types are fed into one of the two silos required for the blend.

The now-blended organic tobacco continues through the primary processing area headed into one of two directions. Some of the tobacco coming out of the conditioning chamber where we added steam and filtered water is cut into strips, then boxed and shipped to Austria and Germany, where contract manufacturers make our tobacco products for sale in Europe.

A part of the primary processing area.

For our domestic markets, where a majority of the tobacco is destined, the organic tobacco is further conditioned, then shredded and fed into a cylinder where it is dried to a moisture level of approximately 13 percent to 14 percent. We then hold this blended tobacco for 24 hours allowing the heat to dissipate before moving on to the next stage.

Guiding SFNTC's primary manufacturing process is a management plan. This covers:

Scope, definitions and acronyms

Policies

Standards

Procedures

Work instructions

Forms, checklists and templates

Tools

Process description

Process inputs

Process outputs

Requirement for outputs

Process activities

Blend strip tobaccos

Produce cut filler

Produce export blended strip

Quality activities

Monitoring strip tobacco moisture

Verifying strip tobacco meets requirements

Monitor cut filler moisture

Verifying cut filler to meet requirements

Monitoring export blended strip moisture

Verifying export blended strip to meet requirements

Flowchart symbols

Process flowchart

Filled cases of organic tobacco are then placed in storage at our warehouses, which, to maintain freshness before shipment to market, are controlled for humidity and temperature.

Once we've created the optimal conditions for cigarette manufacturing—it has the right moisture, temperature, size and shape—the tobacco moves into the *making* area of the building. Here, cigarettes are made under the watchful eyes of our operations people.

## Manufacturing—*Making*

A scene from the manufacturing floor.

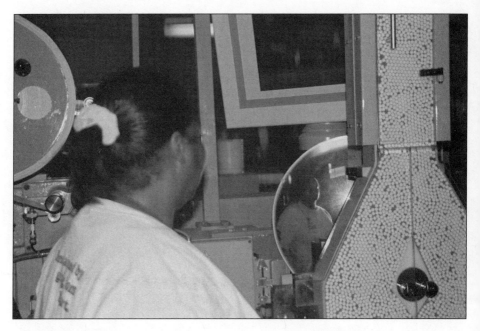

Manufacturing

At the making machines, the tobacco is stored temporarily in hoppers, which continuously allow measured amounts of tobacco to drop down on cigarette paper. Wrapped around the tobacco, the paper is sealed, forming a continuous rod. The rods are then cut to the proper length to make cigarettes.

## SFNTC's Organic System Plan

At every step of processing and manufacturing, organic tobacco must be handled correctly and in accordance with strict National Organic Program standards. It is critically important that organic tobacco not be mixed with any other tobacco, ensuring that it remains free of materials and substances not approved by the USDA National Organic Program.

To help us, we have developed our Organic System Plan. As noted earlier, this is a comprehensive planning and operations guide that includes, among other things, official USDA National Organic Program certification documentation. It contains forms, applications and certification requirements. The plan and its many respective components is not only a guide for those who process and manufacture tobacco products, but for our many suppliers, growers and a host of others who are responsible for one or more aspects of rendering a finished product as well. It is readily available for review by the USDA accredited certifiers and their inspectors.

Our Organic System Plan opens with the actual application filed with Quality Certification Services (QCS), the official organic certifiers. All of our suppliers—including growers—must complete similar applications. Copies of each supplier's certified organic certificates are kept in the plan.

Following are some of the specific requirements we, as an organic tobacco manufacturer, must attend to as we produce certified organic tobacco products.

## The Application

The first order of business is for the manufacturer to apply for organic certification. In our application, we describe who we are and the scope of our business. We are a primary processor, not a contract processor. We cure, dry, mix, separate and clean tobacco, and manufacture tobacco products. Our plant is located at 3220 Knotts Grove Road in Oxford, North Carolina. It is a certified organic facility.

We market our products in the United States, but our tobacco is also exported to Japan, Australia, Canada, Switzerland and the European Union, therefore our organic tobacco must also meet their organic requirements.

We sell our organic tobacco products at wholesale and retail in two cigarette styles and a roll-your-own product. Each is labeled under the USDA organic program as "Organic" or "Made with Organic Tobacco."

We provide a complete written description and schematic product flow charts that show the movement of all organic products from receiving through production and shipping.

### Receiving Stations and Warehouses

SFNTC's general buying and receiving stations and warehouses are facilities that are certified to buy, receive and/or warehouse certified organic tobacco. They do not process or change the physical state of the organic tobacco as received from the organic producers and/or certified organic production facilities. A complete audit trail records the entire process.

## Assurance of Organic Integrity

A major portion of our Organic System Plan includes our *organic integrity program*, which we have put in place to ensure handling practices and procedures present no contamination risk to organic products from commingling with non-organic products or with prohibited substances. For example, packaging materials, bins and storage containers must not have contained synthetic fungicides, preservatives or fumigants. Reusable bags or containers are cleaned and pose no risk to the integrity of organic products. All procedures we employ to maintain organic integrity must be fully documented and in accordance with requirements of the USDA National Organic Program.

## Process-Monitoring Programs

We have process-monitoring programs in place to address areas of potential commingling and contamination. The company also has a quality assurance program in place and is ISO certified. ISO stands for the International Organization for Standardization, the international standard-setting body composed of representatives from various national standards organizations. ISO promulgates world-wide proprietary industrial and commercial standards. Achieving ISO certification is a key achievement.

As part of our monitoring program, we test our tobacco

all the way through the supply chain:

> Prior to purchase
> Upon receipt
> During production
> As finished products

## Equipment

All equipment we use in processing organic tobacco is cleaned and purged thoroughly prior to organic production. Again, this process is fully documented. Equipment we use in processing includes:

> Blending/slicer
> Conditioning cylinder
> Counterflow separator
> Silo units/order cylinders
> Dryers/Cooling and pack
> Cigarette maker/poucher

## Sanitation

The documented cleaning that we perform in the plant in conjunction with our organic program includes sweeping, scraping or vacuuming of the receiving area, tobacco storage, product transfer and production areas, equipment, packaging area, finished product storage, loading dock and building exterior.

All equipment is thoroughly cleaned prior to processing organic tobacco.

## Storage

SFNTC keeps its organic tobacco in dedicated areas and at different stages, including in-process, finished product and off-site storage, all of which are subject to inspection by our certifiers and the USDA National Organic Program.

## Transportation

As in other areas, we carefully manage the transportation of organic tobacco. Incoming tobacco is transported by contract tobacco handlers. Transport companies understand organic handling requirements and the trucks are cleaned prior to transport. As in other areas, the entire process is fully documented. Separate pallets are clearly identified "organic." We handle the tobacco in a separate area in the transport unit. It is sealed in impermeable containers and shrink-wrapped.

We keep records of all purchasing orders, contracts, invoices, receipts, bills of lading, customs forms, scale tickets, quality test results, certificates of analysis, receiving records and receiving summary logs.

When we process our organic tobacco, all transport units have been cleaned. Again, the inspection and cleaning process is fully documented. This documentation includes equipment clean-out logs, ingredient inspection forms and blending, production and packaging reports.

Transport companies are informed of organic handling requirements, which are documented, including shipping logs, clean-truck affidavits, bills of lading, export declaration forms and shipping summary logs.

Additionally, our record-keeping system can track the finished product back to the organic tobacco used in the manufacturing of any of our organic tobacco products.

**Record Keeping**

We maintain all of our organic records for at least five years, and we disclose all activities and transactions of operations involving organic tobacco. All these records are either sent or made available to the designated representative of the USDA's National Organic Program for inspection and copying.

These records include:

Organic certificates for subcontracted operations

Proposed labels for all finished products

Certificates from each supplier of organic tobacco

Regulations for any non-retail containers intended for export labeled in accordance with any requirements of the foreign country of destination or foreign contract buyer for transporting (air or ship) product for export

A complete written description or schematic product flow chart that shows the movement of all organic products, from incoming/receiving, through production and to outgoing/shipping

Organic profile sheets for each by-product

Water tests

Material safety data sheets and/or product label for boiler additives

Facility maps showing the location of traps and monitors, and submit material safety data sheets and/or label information for substances used for pest control

Facility pest management plans to address "Facility Pest Management Practice Standard" and "Commingling and Contact with Prohibited Substance Prevention Practice Standard"

Organic integrity program, or a list of specific control points (see Control Points heading near the end of this chapter) we have identified in our process, and stating how we have addressed each control point to protect organic integrity

Material safety data sheets and/or product label for each cleanser and sanitizer product used to clean and sanitize surfaces in contact with organic products

All incoming tobacco records, in-process records and outgoing (shipping) records

Organic Standard Operating Procedures guide every part of the organic process

**Export**

Our 100 percent organic strip tobacco blends and our packaged tobacco products made with 100 percent organic tobacco are made in accordance with our certified organic handling processor plan, which is part of our Organic System Plan. For each product exported, SFNTC completes an export affidavit, which is reviewed and signed by organic inspectors prior to export.

Organic packaged tobacco products are also exported under international organic guidelines.

As a producer of organically certified goods for export to the European Union (EU), we must complete an EU Compliance

Plan for our official organic certifiers, Quality Certification Services, to verify compliance with Regulation (EEC) No. 2092/91, which is the legal basis for production, processing and trade of organic products in the 25 countries of the European Union. Only products certified according to this regulation can be labeled as "organic" in the EU.

Another certification carries the EEC 2092/91 (European Union) compliance statement.

This statement notes that the certifier (in this case Quality Certification Services) is a USDA National Organic Program third-party certification program that has demonstrated compliance with the International Organization for Standardization Guide 65 for competency and reliability through completion of an accreditation audit by the USDA Agricultural Marketing Service for the verification and application of the EU standards.

The statement notes that the certifier has determined, based on a review of the grower's application and records, and inspections of its fields, facilities and processes, that the named entity meets or exceeds the appropriate and applicable standards of organic production, handling and processing as established in EEC 2092/91. In displaying this certificate, the certified entity warrants that the grower is in and will remain in full compliance with the EU standard. It, too, like the U.S. certificate, provides an identification number and effective dates.

**Certification for Handling Raw Organic Tobacco**

All organic certification certificates are received and

managed at the SFNTC manufacturing offices in Oxford, North Carolina.

A Producers' List is provided to the contract handler for check-off verification prior to receiving any organic tobacco for processing. The organic tobacco is processed once or twice annually.

## Control Points

*While we've provided some of the following control-point information earlier, it is worth repeating some of those key points and including some new points.*

## Organic Tobacco Growers

All organic tobacco farmers must follow the specific production and quality requirements of the SFNTC certified organic tobacco production program and contract. Each producer must be certified organic by a USDA accredited certifier who conducts an annual inspection of their organic farm and farm system plan. Further, a copy of the organic certification certificate must be furnished to SFNTC with their annual contract renewal.

The certificate each farmer earns and presents to SFNTC comes from Quality Certification Services (QCS) certifying that the grower meets the strict standards to be *Certified Organic.*

The certificate states that QCS, as part of the USDA National Organic Program and ISO-65 compliant organic certification program, has determined that the grower, based on a review of the grower's application and records, and inspection of its fields, facilities and processes, meets or exceeds the appropriate and applicable standards of organic production, handling and processing. In displaying this certificate, the Certified Entity warrants that it is in full compliance with the organic standards set by the USDA National Organic Program.

Each certificate carries an identification number and effective date. It also lists the acreage and crops (organic tobacco and tobacco seedlings, as appropriate).

### Outside Receiving and Storage Contract Handlers

When receiving and buying organic tobacco from our growers by one of our outside handlers, each receipt must be verified by organic producer, contract/registration numbers, weight slip and grade and valet certification certificate and producer list check-off. Prior to receiving and handling organic tobacco, our outside handlers must follow our organic standard operating procedures; clean down all work areas including the receiving area, handling equipment, scale and conveyor system and inventory/storage area. Further, the standard operating procedure and check-off-sheet must be followed and stored in a dedicated area.

### Outside Stemmery Contract Handlers

Upon receiving SFNTC organic tobacco, the handler

must verify the organic tobacco supply, producer, weight slip, grade and lot number. Prior to receiving, handling and processing green leaf tobacco, the handler must carefully follow the standard operating procedures' clean down procedure and check-off. All production equipment scheduled for the organic tobacco batch run must be cleaned. Each section supervisor must verify that the standard operating procedure was performed by completing the check-off-sheet, sign and date. All required organic production records, including final organic product lot numbers and yield, must be carefully maintained.

## SFNTC's Blending Handlers

When we receive organic tobacco at our own facility from our growers, like our outside handlers, we carefully verify the organic supply source, producers, contract/registration numbers, weight and grade. Prior to an organic batch run, our handlers must carefully follow the standard operating clean down procedure and check-off and thoroughly clean down the staging area and all production equipment scheduled for the organic batch run. Each section supervisor must verify that the standard operating procedure was performed by completing the check-off-sheet. Our handlers must carefully maintain all required organic production records, including final organic product lot numbers and yields.

## SFNTC's Organic Pouch Tobacco Staff

Our staff must verify organic tobacco source by lot number and production schedule. Prior to an organic product batch run, members of our staff are required to carefully follow standard operating procedures and clean down/prep and check-off procedures, making sure staging areas are cleaned, using only dedicated/marked handling units. Work stations are thoroughly hand cleaned and organic packaging materials verified. It is essential that production records are maintained and all organic product is lot numbered and production units posted; for example pouch, pouch display and packing carton labeled.

## SFNTC's Tobacco Cigarette-Maker Handlers

Prior to commencing an organic tobacco product cigarette run, SFNTC's maker handlers carefully follow the standard operating procedures clean down procedure and check-off, thoroughly clean down the organic tobacco staging area, cigarette maker unit, packaging and final packing areas. Bulk containers and cigarette holding trays are marked "organic" including all handling equipment and receiving areas.

Types of organic tobacco cigarette (regular, light, ultra light, etc.), gross tobacco in and net final packaged units by type, packaging and lot number are carefully recorded.

## SFNTC's Organic Fulfillment/Bonded Storage Handlers

Upon receiving finished packaged organic tobacco products, SFNTC's storage handlers store product in a clearly identified holding area pending final preparation and shipment. Assurance is made on all shipping documents list, SKU, quantity and lot numbers.

## Internal SFNTC and System Wide Mock Organic Inspections

As part of the overall companywide quality assurance and organic integrity programs, SFNTC staff conduct mock organic inspections of its entire organic tobacco products production system, using the SFNTC recall system protocols.

The mock organic inspections are conducted once or twice annually during critical organic production runs. All production entities are included and participate in the recall process, from farm through finished and shipped product.

After each mock inspection, a team leader files a report with the organic program manager for review and corrective action if required. A copy of this report and the corrective action taken is made available to the organic inspector during the annual inspection.

SFNTC has established a complete audit trail system from purchased organic tobacco through final packaged and shipped organic tobacco products by lot number system, bill of lading and customer invoice. Detailed records are maintained for all standard operating procedures and check off sheets used to prevent the possible contamination and/or commingling of the organic products.

## Non-Tobacco Issues

SFNTC also provides to certifiers information pertaining to non-organic ingredients, such as the cigarette filter, and processing aids. This includes ingredient and processing aids, spec sheets, material safety data sheet forms and non-ingredient attachments. This also includes declarations and documentation from each supplier.

For those outside suppliers who process organic tobacco, SFNTC provides a booklet listing our requirements in processing our tobacco. It is our intent to address most issues that may occur in processing our tobacco and give processors clear and specific instruction on how to meet our expectations, but the company also encourages close communications should any questions arise. The plan details the entire process and requirements, guidelines and quality standards.

# United States Department of Agriculture—National Organic Program

*T*he United States Department of Agriculture (USDA) is the government entity that regulates organic farming, processing and manufacturing in America. In the following pages, we briefly outline why there are national organic standards and USDA's role, as well as provide the USDA's perspective on production, handling and certification. We'll also look at the role of certifying agents, organic labeling and marketing. Farmers began developing organic farming systems in the United States more than 60 years ago. And while the growth and popularity of going organic proceeded slowly, today, organic markets are growing rapidly.

The U.S. Congress passed the Organic Foods Production Act of 1990. The Act required the USDA to develop national standards for organically produced agricultural products to assure consumers that agricultural products marketed as organic meet consistent, uniform standards. The 1990 act resulted in regulations requiring that agricultural products labeled as organic originate from farms or handling operations certified by a state or private entity that has been accredited by USDA.

In 2001, the USDA's National Organic Program was born. The program was the result of many years of hard work and the continuing work of the organic industry, producers, manufacturers, professional groups, government and, most importantly, the consumers of organic products.

The administration of the National Organic Program is the responsibility of USDA's Agricultural Marketing Service. Mandated input and guidance is managed through the National Organic Standards Board—a diversified representative advisory panel composed of 15 members appointed by the Secretary of Agriculture, a member of the President's cabinet.

The uniform standards of the USDA's National Organic Program have facilitated further growth in the organic farm sector, which used to be regulated separately by each state with no uniform set of rules. The standards have leveled the playing field and accelerated growth. USDA's organic standards incorporate cultural, biological, and mechanical practices that foster cycling of resources, ecological balance, and protection of biodiversity—practices that have evolved over the last half-century. "The National Organic Program regulations are flexible enough to accommodate the wide range of operations and products grown and raised in every region of the United States," USDA says.

"An increasing number of U.S. farmers are adopting these systems to lower input costs, conserve nonrenewable resources, capture high-value markets and boost farm income," the USDA says. Despite the time, costs and effort required to meet these stringent requirements, USDA estimates that farmers and ranchers added more than a million acres of certified organic land for major crops and pasture between 1995 and 2003, doubling organic pasture and more than doubling organic cropland for major crops.

In 2003, total certified organic acreage dedicated to all organic crops and pasture amounted to some 2.2 million acres. By 2008, that has grown to 3.7 million acres and the total U.S. organic market now exceeds $20 billion annually.

"Organic farming systems," the USDA says, "rely on ecologically based practices, such as biological pest management and composting; virtually exclude the use of synthetic chemicals, antibiotics and hormones in crop production; and prohibit the use of antibiotics and hormones in livestock production. Under organic farming systems, the fundamental components and natural processes of ecosystems—such as soil organism activities, nutrient cycling, and species distribution and competition—are used as farm management tools."

The national organic standards address the methods, practices and substances used in producing and handling crops.

"Although consumer demand for organic foods is expected to continue growing rapidly in the United States and other major markets, the competition for these markets is likely to increase considerably," USDA says. Since 2002, USDA has accredited more than 40 organizations in foreign countries, as well as approximately 50 groups in the United States.

## USDA Organic Crop Standards

Production and handling standards address organic crop production, wild crop harvesting, organic livestock management and processing, and handling of organic agricultural products. The organic crop production standards require that:

Land must have no prohibited substances applied to it for at least three years before the harvest of an organic crop.

Soil fertility and crop nutrients will be managed through tillage and cultivation practices, crop rotations and cover crops, supplemented with animal and crop waste materials and allowed synthetic materials.

Crop pests, weeds and diseases will be controlled primarily through management practices including physical, mechanical and biological controls. When these practices are not sufficient, a biological, botanical or synthetic substance approved for use on the National List of Allowed Synthetic and Prohibited Non-Synthetic Substances may be used.

Preference will be given to the use of organic seeds and other planting stock, but a farmer may use non-organic seeds and planting stock under specified conditions.

The use of genetic engineering (included in excluded methods), ionizing radiation and sewage sludge is prohibited.

## USDA Organic Handling Standards

The handling standards require:

All non-agricultural ingredients, whether synthetic or non-synthetic, must be included in the National List of Allowed Synthetic and Prohibited Non-Synthetic Substances.

Handlers must prevent the commingling of organic with non-organic products and protect organic products from contact with prohibited substances.

In a processed product labeled as "organic," all agricultural ingredients must be organically produced, unless the ingredient or ingredients are not commercially available in organic form.

Organic crops are raised without using most conventional pesticides, petroleum-based fertilizers or sewage sludge-based fertilizers. Animals raised on an organic operation must be fed organic feed and given access to the outdoors. They are given no antibiotics or growth hormones.

## USDA Labeling Standards

Labeling standards are based on the percentage of organic ingredients in a product. Products labeled "100 percent organic" must contain only organically produced ingredients. Products labeled "organic" must consist of at least 95 percent organically produced ingredients. Products meeting the requirements for "100 percent organic" and "organic" may display the USDA Organic seal.

Processed products that contain at least 70 percent organic ingredients can use the phrase "made with organic ingredients" and list up to three of the organic ingredients or food groups on the principal display panel. For example, soup made with at least 70 percent organic ingredients and only organic vegetables may be labeled either "made with organic peas, potatoes, and carrots," or "made with organic vegetables." The USDA Organic seal cannot be used anywhere on the package.

Processed products that contain less than 70 percent organic ingredients cannot use the term "organic" other than to identify the specific ingredients that are organically produced in the ingredients statement.

"A civil penalty of up to $10,000 for each offense can be levied on any person who knowingly sells or labels as organic a product that is not produced and handled in accordance with the National Organic Program's regulations," USDA says.

**USDA Certification Standards**

The USDA accredits state, private and foreign organizations or persons to become "certifying agents." Certifying agents certify that organic production and handling practices meet the national standards.

**Who or What Needs to be Certified?**

Operations or portions of operations that produce or handle agricultural products that are intended to be sold, labeled or represented as "100 percent organic," "organic," or "made with organic ingredients" or food group(s) need to be certified.

**How Do Farmers and Handlers Become Certified?**

An applicant must submit specific information to an accredited certifying agent. Information must include:

The type of operation to be certified

A history of substances applied to the land for the previous three years

The organic products being grown, raised or processed

The Organic System Plan—a plan describing practices and substances used in production. The plan also must describe monitoring practices to be performed to verify that the plan is effectively implemented, a record-keeping system and practices to prevent commingling of organic and nonorganic products and to prevent contact of products with prohibited substances.

Applicants for certification must keep accurate post-certification records for five years concerning the production, harvesting and handling of agricultural products that are to be sold as organic.

These records must document that the operation is in compliance with the regulations and verify the information provided to the certifying agent. Access to these records must be provided to authorized representatives of USDA, including the certifying agent.

Producers and handling (processing) operations that sell less than $5,000 a year in organic agricultural products are exempt from certification, however, these products cannot be used by commercial organic manufacturers. They may label their products organic if they abide by the standards, but they cannot display the USDA Organic seal. Retail operations, such as grocery stores and restaurants, do not have to be certified.

## USDA Organic Accreditation Standards

Accreditation standards establish the requirements an applicant must meet in order to become a USDA-accredited certifying agent. The standards are designed to ensure that all organic certifying agents act consistently and impartially. Successful applicants will employ experienced personnel, demonstrate their expertise in certifying organic producers and handlers, and prevent conflicts of interest and maintain strict confidentiality. Re-certification is required every five years.

Imported agricultural products may be sold in the United States if they are certified by USDA-accredited certifying agents. Imported products must meet the National Organic Program standards. USDA has accredited certifying agents in several foreign countries.

According to the USDA, "In lieu of USDA accreditation, a foreign entity also may be accredited when USDA 'recognizes' that its government is able to assess and accredit certifying agents as meeting the requirements of the National Organic Program—called a recognition agreement."

## Who is Affected by the Standards?

Any farm, wild crop harvesting or handling operation that wants to sell an agricultural product as organically produced must adhere to the national organic standards. Handling operations include processors and manufacturers of organic products. These requirements include operating under an organic system plan approved by a certifying agent and only using materials in accordance with the National List of Allowed and Prohibited Substances. Operations that sell less than $5,000 a year in organic agricultural products are exempted from certification and preparing an organic system plan, but they must operate in compliance with these regulations in order to label products as organic. Retail food establishments that sell organically produced agricultural products do not need to be certified, although many are certified handlers.

The only exception at this time is the production of fish and seafood. Until the NOP develops standards for fish and seafood, these operations may be certified to other private standards.

## USDA Organic Inspection and Certification Process

Certifying agents review applications for certification eligibility. A qualified inspector conducts an on-site inspection of the applicant's operation. Inspections are scheduled when the inspector can observe the practices used to produce or handle organic products and talk to someone knowledgeable about the operation.

The certifying agent reviews the information submitted by the applicant and the inspector's report. If this information demonstrates that the applicant is complying with the relevant standards and requirements, the certifying agent grants certification and issues a certificate. Certification remains in effect until terminated, either voluntarily or through the enforcement process.

Annual inspections are conducted of each certified operation, and updates of information are provided annually to the certifying agent in advance of conducting these inspections. Certifying agents must be notified by a producer or handler immediately of any changes affecting an operation's compliance with the regulations, such as application of a prohibited pesticide to a field.

## Compliance Review and Enforcement Measures

The regulations permit USDA or the certifying agent to conduct unannounced inspections at any time to adequately enforce the regulations. Certifying agents and USDA may also conduct pre- or post-harvest testing if there is reason to believe that an agricultural input or product has come into contact with a prohibited substance or been produced using an excluded method.

## Organic Agriculture

Official USDA definition of organic production: "A production system that is managed in accordance with the Organic Foods Production Act and regulations to respond to site-specific conditions by integrating cultural, biological and mechanical practices that foster cycling of resources, promote ecological balance and conserve biodiversity."

## National Organic Program

Official USDA definition: The National Organic Program is USDA's Agricultural Marketing Service. It administers Federal regulations on organic standards and certification.

# Resources for the Organic Tobacco Grower

**Santa Fe Natural Tobacco Company**

**Leaf Department (The Tobacco Growers' Primary Contact)**
Fielding Daniel, Director
Randal Ball, Manager
105 West Street
Oxford, North Carolina 27565
919-690-1905
Manufacturing Operations
3220 Knotts Grove Road
Oxford, North Carolina 27565
919-690-0880
Company Offices/Headquarters/Administration
One Plaza La Prensa
Santa Fe, New Mexico 87507
505-982-4257
www.sfntc.com

## Carolina Farm Stewardship Association

The Carolina Farm Stewardship Association (CFSA) was established in 1979 and is a farmer-driven, non-profit organization serving North and South Carolina with a mission of inspiring, educating and organizing farmers, consumers and businesses in the Carolinas to develop a sustainable regional food system that is good for farmers and farm workers, good for consumers and good for the environment.

CFSA developed the first organic certification standards for farms in the Carolinas, and administered those standards until the establishment of the USDA's National Organic Program. The organization functions as a network for individuals and organizations with an interest in sustainable food systems. Activities include:

Educational and networking opportunities for farmers and food activists, such as its annual Sustainable Agriculture Conference, which is the largest multi-day sustainable agriculture conference in the Southeast Atlantic Seaboard foodshed.

Consumer outreach efforts such as its regional farm tours, the largest farm tour program in the country.

Infrastructure development projects that promote access to regional wholesale markets for farmers, such as its establishment of Eastern Carolina Organics, which is now an independent, highly successful, farmer-owned for-profit wholesale distributor of North Carolina grown organic produce.

Community food system development projects, such as its Farm Incubator Program, which is helping communities throughout the Carolinas establish mentoring farms where people without farming backgrounds can gain the real world production and marketing experiences necessary to successfully launch their own farming enterprises.

Contact:
Carolina Farm Stewardship Association
PO Box 448
Pittsboro, North Carolina 27312
919-542-2402 phone
919-542-7401 fax
www.carolinafarmstewards.org

## Agrisystems International
## SFNTC's Organic Program Consultant

Agrisystems International—business consultants specializing in organic production systems development for farm, manufacturers and trade channels. Assists producers in preparing for organic systems inspection and certification. Also provides document preparation and submission, as well as label, ingredient or materials and audit trail review. Offers on-site evaluation and training. Agrisystems International provides organic consulting services to SFNTC.

Contact:
Thomas B. Harding, Jr.
President
Agrisystems International
125 West 7<sup>th</sup> Street
Wind Gap, Pennsylvania 18091
610-863-6700 phone
610-863-4622 fax

**Cooperative Extension Service**

A national network of land-grant universities and the U.S. Department of Agriculture. *(See each state's extension services. Listed here is information on North Carolina's program.)*

North Carolina's Cooperative Extension Service is based in the College of Agriculture and Life Sciences and provides access to farmers and the public seeking information and expertise through local centers in the state's 100 counties and on the Cherokee Reservation. It is part of a national network of land-grant universities including North Carolina State University and North Carolina Agriculture &Technical State University, and the U.S. Department of Agriculture.

In North Carolina and nationally, Cooperative Extension provides educational programming in five areas:

Sustaining agriculture and forestry

Protecting the environment
Maintaining viable communities
Developing responsible youth
Developing strong, healthy and safe families

As in many other states, in county centers across the state, county agents are a contact point for growers and extension specialists working at the state's land-grant universities. Agents educate the public through meetings and workshops, field days, personal consultations and satellite broadcasts. They also provide publications, newsletters, computer programs, videotapes and other educational materials.

Contact:
Consult Your County Extension Service Agent, or
North Carolina State University
Agriculture & Resources Economics
North Carolina Cooperative Extension Service
Box 8109
Raleigh, North Carolina 27695-8109
919-515-3107 phone
919-515-6268 fax
www.ag-econ.ncsu.edu/extension
North Carolina Agriculture &Technical State University
1601 East Market Street
Greensboro, North Carolina 27411
336-334-7500 phone
www.ncat.edu

## Quality Certification Services (QCS)

QCS offers organic certification accredited by the USDA National Organic Program and USDA ISO Guide 65. QCS certifies organic tobacco farming. It also certifies wildcrafting (harvesting plants from their natural, or wild habitat), livestock, processing, packing and handling operations. QCS certifies a diverse array of organic operations regardless of type, location or size.

### Overview of the Organic Certification Process

When an operation decides to seek Organic Certification with QCS, the first step in the certification process is to order a Certification Application Packet. This packet contains the QCS Certification Handbook, the current Organic Materials Review Institute manual of generic materials allowed, restricted and prohibited for use in organic production, and the application(s) for certification. Certification Application Packets are available for a one-time fee of $25, and can be ordered by mailing the QCS Office a written request for the type of packet (Farm, Livestock, or Processor-Handler-Packer) along with a check or money order for $25.

Once the completed application, along with supporting documentation and relevant certification fees, arrives at the QCS office, the certification process normally takes anywhere from six to 12 weeks.

First, the application is reviewed by a QCS certification coordinator for completeness and basic compliance to organic standards. If necessary, the applicant may be contacted to provide additional information or documentation.

Next, the application is forwarded to an independent organic inspector, who will then contact the applicant to schedule the inspection of the farm or facility.

Once the inspection is complete and the inspector submits the inspection report to QCS, the application then enters the final stage of the Organic Certification process. QCS certification coordinators again review the application, gathering information as necessary, and make a certification decision. If approved, the certificate is then issued to the applicant.

Contact:
Quality Certification Services
PO Box 12311
Gainesville, Florida 32604
352-377-0133
www.qcsinfo.org

## United States Department of Agriculture (USDA)

See Chapter 8—USDA and the National Organic Program—for more detailed information. USDA provides leadership on food, agriculture, natural resources and related issues based on sound public policy, the best available science and efficient management. USDA strives to provide the integrated program needed to lead a rapidly evolving food and agriculture system.

USDA's National Organic Program develops, implements, and administers national production, handling and labeling standards for organic agricultural products. The National Organic Plan also accredits the certifying agents (foreign and domestic) who inspect organic production and handling operations to certify that they meet USDA standards.

Contact:
U.S. Department of Agriculture
1400 Independence Ave., S.W.
Washington, D.C. 20250
www.usda.gov/AMS/NOP

## ATTRA—the National Sustainable Agriculture Information Service

ATTRA, the National Sustainable Agriculture Information Service, is managed by the National Center for Appropriate Technology and is funded under a grant from the United States Department of Agriculture's Rural Business-Cooperative Service. Visit the NCAT Web site, noted below, for more information on sustainable agriculture projects.

Whether you are a beginning farmer, or an experienced agricultural producer interested in transitioning to more sustainable practices, you may want to start by learning more about the principles of sustainable agriculture and some of the systems approaches associated with it. The publications in this series introduce and discuss concepts, and provide a general overview of planning and managing a more sustainable farm operation.

The site contains the latest in sustainable agriculture and organic farming news, events and funding opportunities. It includes in-depth publications on production practices, alternative crop and livestock enterprises, innovative marketing, organic certification and highlights of local, regional, USDA and other federal sustainable agricultural activities.

Contact:
ATTRA—National Sustainable Agriculture Information Service
PO Box 3657
Fayetteville, Arkansas 72702
800-346-9140
www.attra.ncat.org

## North Carolina Department of Agriculture and Consumer Services
## Agronomics Division

The division's mission is to provide North Carolinians with diagnostic and advisory services that increase agricultural productivity, promote responsible land management and safeguard environmental quality.

Regional agronomists provide on-site consulting services to help growers troubleshoot nutrient and nematode problems, establish appropriate agronomic sampling programs and implement management recommendations in a cost-effective and environmentally sound manner.

North Carolina Department of Agriculture's agronomist Robin Watson provides hands-on advice to an organic tobacco grower.

Thirteen regional agronomists within the Agronomic Division's field services section help growers manage all types of crop-nutrient issues. They make site visits, provide advice on lime and fertilizer application, demonstrate sampling techniques and help growers troubleshoot and manage potential nutrient problems. The NCDA&CS Field Services section has been helping North Carolina growers manage fertilization and other nutrient-related issues for nearly 30 years.

Contact:
North Carolina Department of Agriculture and Consumer Services
1040 Mail Service Center
Raleigh, North Carolina 27699-1040
919-733-2655 phone
919-733-2837 fax
www.ncagr.com/agronomi/rahome.htm

## Virginia Department of Agriculture and Consumer Services

The department's mission is to foster a sound agricultural economy and to help create jobs and economic investment in Virginia agribusiness. To accomplish this mission, the department is recruiting agricultural businesses to Virginia and facilitating the expansion of existing agricultural industry.

Established in 1877, the department is responsible for over 60 laws and more than 70 regulations relating to consumer protection and the promotion of agriculture. The department is located within the Governor's Secretariat of Agriculture and Forestry and has both economic development and regulatory responsibilities under state law.

Agency employees work in and out of the department's several field offices; in the agency's five regional laboratories located in Warrenton, Lynchburg, Ivor, Harrisonburg and Wytheville; at an international office in Hong Kong and at its headquarters in Richmond.

The department works in cooperation with Virginia State

University, Virginia Tech and the Virginia Cooperative Extension Service on research, education and marketing projects.

Contact:
Virginia Department of Agriculture and Consumer Services
102 Governor Street
Richmond, Virginia 23219
804-786-2373
www.vdacs.virginia.gov

## South Carolina Department of Agriculture

The department's mission is to promote and nurture the growth and development of South Carolina's agriculture industry and its related businesses while assuring the safety and security of the buying public.

The Agricultural Services Division consists of Agribusiness Development, Marketing and Promotion, State Farmers Markets, Market News Service and the Grading and Inspection Program.

According to the department, the South Carolina Department of Agriculture's Small Farms Program was the first of its kind in the country. The program provides assistance to small family farmers with an emphasis on dissemination of information, referrals and counseling on issues such as land retention, alternative land use and community development. The focus of the Small Farms Program is to assist small farmers in understanding the challenges associated with retail marketing and in helping them to find solutions to their specific problems.

The department works with several governmental agencies and non-profit partners to provide training materials to small farmers and market managers to help them develop and or improve their marketing skills.

The National Commission on Small Farms defined small farms as farms with sales less than $250,000 annually. The commission wanted to include more farm families of relatively modest means who may need to improve their net farm incomes. According to the 2002 agricultural census, 96 percent of all farms in South Carolina are small farms.

The department provides information resources to farmers interested in marketing their organic crops. South Carolina organic farmers currently market their crops at farmers markets, community supported agriculture farms, specialty shops, farm stands and wholesale distributors.

Other South Carolina Resources:

Clemson University Organic Certification Program
Certified Organic Operations in South Carolina

Contact:
South Carolina Department of Agriculture
PO Box 11280
Columbia, South Carolina 29211
803-806-3820
www.agriculture.sc.gov

### Kentucky Department of Agriculture
### Office of Agriculture Marketing and Product Promotion

The department serves Kentucky farmers looking to enhance or develop markets for a variety of horticultural crops. Marketing specialists serve farmers' markets, fruit and vegetable growers, ornamental agriculture producers and certified organic farmers.

Agricultural education strives to improve agriculture literacy by developing programs that increase consumer, educator and student awareness about agriculture. Farm safety strives to increase safety awareness and provide educational resources and training to ensure the safety of farmers and their families. Farmland preservation allows the state to purchase agriculture conservation easements to ensure that lands currently in agricultural use will continue to remain available for agriculture and will not be converted to other uses.

The Division of Agriculture Marketing and Agribusiness Recruitment works with agriculture parties showing an interest in locating in Kentucky. While working with these companies the division assists in property location, financing and tax incentive programs. The division also works with businesses wishing to expand by helping them find financial assistance.

The technical support branch develops pesticide environmental strategies and programs that are designed to protect the land and water of the Commonwealth from agriculture pesticide use through best management practices.

Contact:
Kentucky Department of Agriculture
32 Fountain Place
Frankfort, Kentucky 40601
502-564-4696 phone
502-564-2133 fax
www.kyagr.com

## Tennessee Department of Agriculture
## Tennessee Agricultural Enhancement Program

The program was developed to promote long-term investments in Tennessee's livestock and farming operations by providing cost-share funds to qualifying producers. The enhancement program allows producers to maximize farm profits, adapt to current market situations and prepare for the future. In addition, the program also enables producers to make a positive economic impact in their community. The program is a direct result of the State of Tennessee's continued commitment to supporting farm development and Tennessee's agricultural community.

The Tennessee Department of Agriculture's Agribusiness Development Section exists to provide information and services for prospective or expanding agribusiness. The department will coordinate with other state and local agencies to provide one-stop shopping. In addition, the department can also draw on two other resources—the University of Tennessee's Center for Profitable Agriculture and its Forest Products Center.

The Organic Certification Cost Share Program in Tennessee seeks to defray the cost of organic certification for producers and handlers of organic agricultural products in Tennessee. The Tennessee Department of Agriculture will reimburse each eligible producer or handler for up to 75 percent of organic certification costs, not to exceed $500. Reimbursement will again be available to production and handling operations inspected and certified and/or inspected and receiving continuation of certification as long as funding is available. Producers can also receive re-certification assistance the first year at 50 percent cost-share and the second year at 35 percent.

Contact:
Tennessee Department of Agriculture
PO Box 40627
Nashville, Tennessee 37204
615-837-5160 phone
615-837-5194 fax
www.tennessee.gov/agriculture

**Ohio Department of Agriculture**

The Ohio Department of Agriculture provides regulatory protection to producers, agribusiness and the consuming public; to promote Ohio agricultural products in domestic and international markets; and to educate the citizens of Ohio about the state's agricultural industry.

Contact:
Ohio Department of Agriculture
8995 E. Main St.
Reynoldsburg, Ohio 43068
614-728-6200
www.ohioagriculture.gov

## Ohio Farm Bureau

With more than 230,000 members, the Ohio Farm Bureau is Ohio's largest general farm organization. The Ohio Farm Bureau is a federation of 87 county Farm Bureaus representing all 88 counties.

Farm Bureau members in every county in the state serve on boards and committees working on legislation, regulations and issues that affect agriculture, rural areas and Ohio citizens in general.

Contact:
Ohio Farm Bureau Federation, Inc.
280 Plaza
PO Box 182383
Columbus, Ohio 43218-2383
614-249-2400
www.ofbf.org

# About Santa Fe Natural Tobacco Company

S anta Fe Natural Tobacco Company was started in 1982 in Santa Fe, New Mexico. A couple of friends had a dream of creating a better cigarette. They looked to their surroundings, drawing inspiration from the American Indian culture that is so prevalent in the West and the respectful belief that tobacco should be used in moderation, and in its natural state.

In the early years, the tiny company worked out of a shed in the Santa Fe railroad yards where employees were frequently seen hand-packing tobacco pouches while taking product orders over the telephone. With success and growth, the company moved in 1992 to a former commercial laundry and its outlying buildings.

Four years later, a small manufacturing plant was constructed in Oxford, North Carolina, deep in the heart of tobacco country and close to the company's growing number of growers.

Manufacturing operations in Oxford, North Carolina

In 2002, SFNTC became an independent operating unit of what is now Reynolds American Inc. Rick Sanders, who became president & chief executive officer in 2002, has led the company through unparalleled growth while maintaining the character, values and vision of a truly unique company since then.

To accommodate its steady growth, SFNTC moved its Santa Fe offices to a new adobe-style building in 2004. The building, which has served as a backdrop for our annual holiday card we send to our customers, reflects the character of the company and its surroundings. The building is energy efficient and uses recycled materials.

Main offices in Santa Fe, New Mexico

As with all SFNTC buildings, which include Oxford manufacturing operations and our western distribution center in Reno, Nevada, 100 percent of all electrical power needs are provided by wind-generated energy sources.

To keep connected to its roots and to honor the inspiration for the company, SFNTC established in the mid-1990s the Santa Fe Natural Tobacco Company Foundation. This non-profit organization provides grants to preserve, promote and advance American Indian self-sufficiency, education, language and culture.

SFNTC is made up of talented, diverse, fun loving and sometimes quirky individuals who believe being different is really about being better. Every SFNTC employee is treated fairly, equitably and with respect. The diversity of SFNTC ideals, backgrounds and perspectives is what makes the company what it is today.

Santa Fe-based employees

Santa Fe-based employees reflect the city's history and cultural diversity—American Indian, Hispanic and Anglo. Most of the North Carolina-based employees have deep roots in tobacco, from their own family farms to working for other now long-gone tobacco manufacturers.

Oxford-based employees

If there is one thing all SFNTC employees have in common, it's taking time to break bread together. In Santa Fe, regular company luncheons feature New Mexican fare, with plenty of red and green chili. While in Oxford, employees enjoy regular Pig Pickins', the traditional North Carolina barbecue feast.

Western distribution center employees based in Nevada

Road clean up in Oxford, North Carolina.

While the company doesn't grow tobacco in the high desert of Santa Fe (though it tried), SFNTC promotes earth-friendly practices there as well. In November 2007, the company began a three-year commitment to provide $150,000 to support the Santa Fe Watershed Association's efforts to save the Santa Fe River. The river, which stretches from the Sangre de Cristo Mountains to the Rio Grande—though a majority of the riverbed is dry most of the year—was named "America's Most Endangered River" for 2007 by American Rivers, a Washington, D.C.,-based group.

In addition to providing financial support, Santa Fe employees have adopted a stretch of the river and are providing other volunteer support, such as planting hundreds of willow trees and cleaning up. The Santa Fe Watershed Association's Living River Initiative focuses on raising awareness, public outreach and stakeholder meetings and dialogues to generate support for restoring an environmentally useful flow to the river.

Environmental sustainability efforts stretch into every area of Santa Fe's daily operations. In addition to contracting for 100 percent wind-generated power, in late 2007 the company began ordering new hybrid-powered vehicles to move the entire fleet of vehicles for our sales team to hybrid power. The "silverware" in the employee cafeteria is made of organic biodegradable materials.

These renewable energy purchases by SFNTC eliminate some 2,264 tons of carbon dioxide, 2.39 tons of sulfur dioxide and 3.33 tons of nitrogen oxide each year generated by fossil fuel utility plants. The company also is working to attain ISO 14001 certification, the environmental management component of the International Organization for Standardization. The company will be identifying further efforts to control the environmental impact of our activities and products and continue to improve its environmental performance.

In New Mexico, SFNTC and its employees are working to help restore the Santa Fe River, named one of America's most endangered.

*The People of Santa Fe Natural Tobacco Company*
*share a values driven vision:*
*That our uncompromising commitment to*
*our natural tobacco products,*
*the earth from which they come,*
*the communities on which we depend,*
*and the people who bring our spirit to life,*
*is essential to our success.*

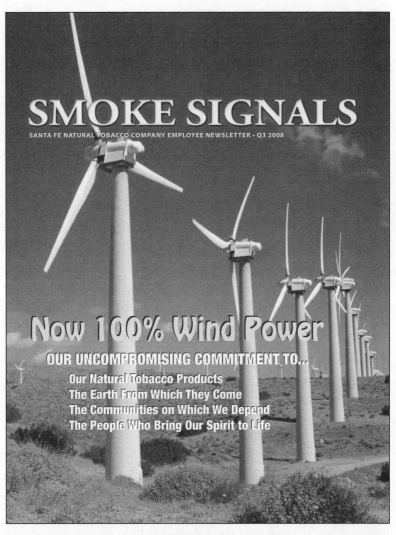

**SMOKE SIGNALS**

SANTA FE NATURAL TOBACCO COMPANY EMPLOYEE NEWSLETTER • Q3 2008

Now 100% Wind Power

OUR UNCOMPROMISING COMMITMENT TO...

Our Natural Tobacco Products
The Earth From Which They Come
The Communities on Which We Depend
The People Who Bring Our Spirit to Life

The company gets all its electricity from wind generated power; front cover Smoke Signals, quarterly employee publication.

Watercolor by artist and SFNTC ISO coordinator John Brassard

SFNTC growers, like Richard Ward, display signs on their farms officially designating them as earth-friendly tobacco growers.

# Acknowledgements

*I*t is with heartfelt thanks that the authors acknowledge the many people who helped make this book possible, especially the many growers who contributed, and those family, friends, and coworkers who provided such valuable support, including Rick Sanders, Haney Bell, Rudy Cook, Gerry Deschenes, John Franzino, Rusty Gaston, Colin Uffindell, Susanne Farr, Sandi Thomas, Cheryl Nizio, Jeanne Dvorak, Alexandra Pratt, John Dillon, Steven Mosher, John Brassard, Randal Ball, Julie Ball, Tom Harding, Robin Sommers, Leigh Park, Chris Webster and the members of North Carolina Department of Agriculture.

This book was not produced with the intention of being a commercial success. Rather it was a labor of love by the People of Santa Fe Natural Tobacco Company to share what we have learned about organic tobacco. Any profits derived from the sale of this book will be contributed to the Carolina Farm Stewardship Association in recognition of the work performed by the association and its members to promote organic and earth-friendly farming.